Praise for *Birding*

"*Birding Under the Influence* is an adrenaline hit for birders and adventure junkies. It's also a surprisingly tender story of redemption, as Dorian Anderson faces down his addictions and reinvents his career. Having eagerly awaited this book after Anderson's 2014 Big Year, I read it in a blissful, all-out binge—as will anyone with a taste for birds and inspiring journeys."

—NOAH STRYCKER, Associate Editor of *Birding* magazine
and author of *Birding Without Borders*

"A story led by the bike and fueled by his tenacity, Dorian colorfully articulates the depth to which putting in the work—both physically on the bike, and emotionally on oneself—reaps infinite rewards."

—TIFFANY KERSTEN, birding guide and
past Lower 48 U.S. Continental Big Year record holder

"What an incredible story! Dorian's adventure is an inspiration for birders and non-birders alike."

—DAVID LINDO, author of *The Urban Birder*

"*Birding Under the Influence* is an incredible story. This has to be the best Big Year book since *Kingbird Highway*. Of course, being a Dutchman, a bicycle Big Year really appealed to me, so I was already looking forward to this book. But this story proved to be so much more. It is not just a cycling journey of 17,800 miles across the US(!), a country that has, unlike the Netherlands, hardly any cycling paths and a lot more relief, but it is a journey through the mind and the soul. It is a coming-of-age story unlike anything I've read before. I know what it takes to do a Big Year—the sacrifices, the willpower needed, the risks involved, and the stress on yourself and your relationships. To do this on a bike and put your academic career behind you while recovering from alcohol and drug addiction is something else. I read this incredible story in one go, and I think you will too."

—ARJAN DWARSHUIS, author of *The (Big) Year That Flew By*

"Dorian is a big personality with an even bigger story. His epic tale of recovery and perseverance will inspire any birder, whether or not their Big Year dreams involve cycling across the continent."

— NATE SWICK, American Birding Association podcast host and author of *The Beginner's Guide to Birding*

"Dorian tells it like it is: don't let life happen to you, no matter what your tendencies and what well-worn grooves you slide into. Life is what you create, accepting your faults and challenges, and realizing that the path forward is not how you plan it—instead, it happens how you least imagined it. And as in any wonderful, gripping story, great birding was involved!"

— ALVARO JARAMILLO, owner and guide, Alvaro's Adventures

"Recounting a starkly different kind of 'Big Year' Dorian details an unlikely saga that takes a toll on him, physically and mentally. So often birding is a quest, and in this memoir Dorian Anderson seeks birds but also some sense of self. At times throwing himself on the mercy of his fellow Americans, he finds his way through and across the country, enlisting a colorful cast of characters as he goes. For this one year his only commitment is to his bicycle. His bike both frees and imprisons him simultaneously, but eventually it delivers him, along with a unique story. America's sweetest wildlife spectacles light a path that would otherwise be strewn with 40-ounce bottles of malt liquor and rails of cocaine. From Snowy Owls in Boston to Yellow-footed Gulls in California's Salton Sea, saddle up for a modern journey that harkens back to classic birding adventures, like those of Pete Dunne in *The Feather Quest* or to *Wild America* by Roger Tory Peterson and James Fisher."

— GEORGE ARMISTEAD, founder and guide, Hillstar Nature

"This is no ordinary Big Year birding book, as Dorian's story supersedes birds and reveals how a year of contemplation, biking, and birding helped him to overcome his inner struggles of addiction and broken relationships that are all too relatable for many of us. Many of us are reluctant to hit the 'pause' button on life, afraid of what that might lead to, but Dorian's example of setting aside a year to process his thoughts, relationships, and future is an example for us, and birding might just be the adventure we need, even for a day or two."

— LUKE SAFFORD, Director of Engagement and Education, Tucson Audubon

Birding Under the Influence

Birding Under the Influence

Cycling Across America
in Search of
Birds and Recovery

DORIAN ANDERSON

Chelsea Green Publishing
White River Junction, Vermont
London, UK

Project Manager: Rebecca Springer
Developmental Editor: Matthew Derr
Copy Editor: Angela Boyle
Proofreader: Nancy A. Crompton
Designer: Melissa Jacobson

Printed in Canada.
First printing November 2023.
10 9 8 7 6 5 4 3 2 1   23 24 25 26 27

Our Commitment to Green Publishing
Chelsea Green sees publishing as a tool for cultural change and ecological stewardship. We strive to align our book manufacturing practices with our editorial mission and to reduce the impact of our business enterprise in the environment. We print our books using vegetable-based inks whenever possible. This book may cost slightly more because it was printed on paper that contains recycled fiber, and we hope you'll agree that it's worth it. *Birding Under the Influence* was printed on paper supplied by Marquis that is made of recycled materials and other controlled sources.

Library of Congress Cataloging-in-Publication Data
Names: Anderson, Dorian, 1978– author.
Title: Birding under the influence : cycling across America in search of birds and recovery /
    Dorian Anderson.
Description: White River Junction, Vermont : Chelsea Green Publishing, [2023]
Identifiers: LCCN 2023030761 | ISBN 9781645022237 (paperback) |
    ISBN 9781645022244 (ebook) | 9781645022251 (audiobook)
Subjects: LCSH: Anderson, Dorian, 1978– | Bird watchers—United States—Bigraphy. |
    Bicycle touring—United States—Bigraphy.
Classification: LCC QL677.5 .A53 2023 | DDC 598.072/34—dc23/eng/20230713
LC record available at https://lccn.loc.gov/2023030761

Chelsea Green Publishing
White River Junction, Vermont, USA
London, UK

www.chelseagreen.com

Sonia, mi amor: gracias por tu paciencia, apoyo,
y compasión. Esta historia es tanto tuya como mía,
y no puedo imaginar dónde estaría sin ti.
Te quiero, ahora y siempre.

Contents

Author's Note

Two helpful notes for readers:

First, bird names are in constant flux because of taxonomy changes and—increasingly—political considerations. Magnificent Hummingbird, which I observed in Arizona, has been split into two species, Rivoli's Hummingbird and Talamanca Hummingbird, based on genetic evidence, and McCown's Longspur, which I observed in Colorado, has been renamed Thick-Billed Longspur because it wasn't advisable to memorialize a Confederate general who also made war on Native Americans. Regardless of why these and the other names were changed, I have used names as the American Ornithologists' Union recognized them in 2014, the year when this story unfolds.

Second, and as you will discover shortly, wind is an important consideration for birders and cyclists. Some readers might not be familiar with how directionality is conveyed, so I'll clarify that a northwest wind is one blowing *from* the northwest (and thereby *toward* the southeast). Likewise, an east wind blows from east to west. I maintain this convention throughout this work.

ONE

Another Time

Boarding the ferry, I questioned my purpose; towering clouds threatened rain, and biting wind taunted as it hissed through stacks of lobster traps. As one of just a few passengers on that November morning in 2012, I sought refuge in the cabin and claimed a window seat, my legs bouncing on the balls of my feet while my finger picked at a crack in the adjacent seat. The rumble of the ship's engines signaled our departure, and my breathing constricted as the heavy craft labored out of the harbor and into the open ocean, the thirty-mile crossing from Hyannis to Nantucket scheduled for two-and-a-half hours.

While the northeastern United States reeled from the devastation that Superstorm Sandy wrought four days earlier, I embarked on a cockamamie quest in the cyclone's aftermath. The prize I sought was the Northern Lapwing, an iridescent, sandpiper-like bird sporting a whimsical black crest. The species ranges through Europe and Asia, so Nantucket birders surprised everyone when they reported two vagrants—representatives that wandered outside the species' usual range—on the island late in October 2012. The pair was likely sucked across the Atlantic by Sandy's gargantuan wind field—it stretched to Scandinavia—and I, a scientific researcher in Boston, hatched plans to travel to Nantucket and view the exotic visitors. If they didn't fly away before I arrived, then they'd be a wonderful addition to my life list, the collection of bird sightings I'd been accumulating since age seven.

Superstorm fallout had prevented earlier travel, but positive reports persisted and buoyed my hopes as I scrambled to complete experiments and clear my calendar. While all birds are beautiful and beguiling, vagrants elicit a particular excitement because no one knows where they'll appear and how long they'll stay. Chasing such transients is an exciting game; they can depart their discovery points at any moment, and the rollercoaster of triumphs and disappointments recalls the sinusoidal cycles of gambling or substance abuse, with dejected birders routinely swearing off chasing vagrants before jumping into the car at the next exciting report. As an alcoholic-addict, I couldn't refuse a rare bird any more than I could a shot of Jägermeister, a line of cocaine, or a hit of ecstasy.

Northern Lapwing possibilities strumming my serotonergic circuits while Nirvana's *Nevermind* blared through my car's speakers, I sped from Boston to Hyannis in the predawn hours of November 2nd. Unfortunately, the ferry didn't share my urgency. The boxy boat pushed through waves as efficiently as a 200-ton Twinkie, and I counted the minutes until our arrival, my fear of departed lapwings swelling each second. Great Point Lighthouse eventually came into view on the port-side horizon, and sweeping dunes rose from the waves as we inched toward the island and into the harbor, where the ship docked. I grabbed my backpack, disembarked, and hustled two blocks to the island's bike shop under strengthening rain. With the local taxi service shut down for the winter and the thought of paying $300 to put my car onto the ferry never having been entertained, my preparatory research had revealed that a bicycle would be my best mode of transport.

Opening the door and stepping inside, I saw a middle-aged man fiddling with a bike. He stood and wiped his greasy hands on a rag.

"Can I help you?" he asked.

I caught my breath and replied, "Yeah, I called yesterday about a rental."

"Oh, you. The bird guy," he said. "I can't believe you came in this weather. Are your *lopwangs* still here?"

I didn't bother correcting his pronunciation. "Hopefully," I said. "They've been seen for the last three days, so there's a good chance."

"Well, let's get you going. Take that one," he said, pointing at a silver-framed hybrid bike. "It rolls great and can handle bumps if you go off-road."

I surrendered my credit card, adjusted the seat, and pushed the rental out the door. Thirty-three years old, I hadn't biked since I graduated college in 2001, so I was curious how the remainder of my pursuit would unfold, especially with twenty-five pounds of gear—binoculars, spotting scope, tripod, camera, and telephoto lens—in my backpack. I mounted up and shoved off, raindrops tapping on my helmet as I wobbled along Broad Street.

Settled in the late seventeenth century, Nantucket grew into a major whaling center in the ensuing decades. The decline of that industry coupled with a massive fire to depopulate the main town by the mid-nineteenth century, and the island sat mostly forgotten until the mid-twentieth, when developers realized it would be an ideal getaway spot for mainlanders. Tourism has sustained the summer sanctuary since, and regulations prohibiting tall buildings and chain restaurants now preserve the island's rustic charm. Gaining confidence as I pedaled past cutesy boutiques, independent eateries, and weathered, gray-shingled cottages, I rolled out of town and into rural surroundings.

Online reports indicated the lapwings moved around the southwestern side of the island, so I hurried toward Hummock Pond, a marshy area where the pair was seen the previous afternoon. Legs burning and chest heaving after four frantic miles, I turned onto a muddy track and sprinted the final 200 yards to a dead end at an elevated overlook. I dismounted, flipped out the kickstand, and shed my backpack. Three days of anticipation converged as I excavated my binoculars. I took a deep breath to steady my shaking hands and lifted the optics to my eyes.

My gaze sweeping along the far shoreline, I seized on two blobs at the base of the reeds. My heart thumped, and I scrambled to mount my spotting scope on my tripod for a better view. Pointing the scope across the water, I spun the focus wheel and brought two dove-sized birds into focus. Bronzy green above and white below, each sported a dark breast band, a buff face, and a black crest, the last feature punctuating the beautiful bird like a chocolate wafer on the top of an ice cream sundae.

Elation replaced angst in that sweetest instant. "Yes! Yes! Yes!" I grunted through clenched teeth. Pumping my fists while stomping my feet, I didn't care that I was splashing mud all over myself. Following two days of forced delay, seventy-five miles in the car, thirty on the boat, and a final four on the bike, this was an unlikely victory, one I'd recount for my birding buddies for as long as I lived.

The rain abated, and my high persisted while I watched and photographed the birds for the next ninety minutes. Lapwings are members of the plover family and grouped with avocets, oystercatchers, sandpipers, godwits, and curlews under the broad umbrella of "shorebirds." Dainty and elegant while probing beaches, mudflats, and fields for invertebrate morsels, shorebirds are powerful fliers; long wings propel them forward at high speed, and some species migrate from the Arctic to the southern hemisphere and back each year. It was that evolved endurance that kept the Nantucket lapwings aloft while Sandy blew them across the Atlantic, presumably over several days.

The birds suddenly took flight. Their black-and-white underwings flashed with each flap, and their squeaky calls carried over the windswept landscape. As I watched the pair turn toward the ocean, I was happy the birds were healthy enough to fly but feared they'd disappear before other birders could enjoy them. The pair doubled back over the dunes, and I relaxed when they landed on a muddy island at the other end of the elongated pond. Figuring I'd had my best views, I broke down my gear, loaded my backpack, and returned to my bicycle under renewed rain.

Lapwing-imperative had forced my outgoing pace, but I rode slower after adding them to my life list. Birds appeared along the roadside despite the dreary conditions, and I paused to appreciate a streaky Song Sparrow belting its trademark tune from the top of a roadside bramble. The melody was motivational, and I smiled when I spotted an overhead Great Blue Heron a few minutes later, the behemoth's rhythmic flaps in phase with my pedal rotations as I cranked along the country lane. In the company of those and other local birds, my return ride to the bike shop was pure joy.

The proprietor greeted me as I burst through the shop door. "How'd it go?" he asked.

"I got 'em!" I bellowed.

"Great! Woulda been a long ferry ride home without 'em!" he said.

"I know. I missed the Gray-tailed Tattler, another Eurasian shore-bird, that was here on October eighteenth."

His eyes opened wide. "This is your second trip from Boston in two weeks. For birds?" he asked disbelievingly.

"You know it! I didn't need the bike last time because the tattler was in the harbor. At least, it was until I got here. Just missed it. But that's how this treasure hunt goes. Find 'em or miss 'em, I'm happy to be outdoors, looking at whatever birds I can find."

My departure time was nearing, so I thanked him for the rental and started toward the ferry. I boarded and flopped into the same seat I occupied on the earlier crossing, my exhausted legs as still as fence posts as the boat pulled away from the dock.

Nantucket fading into the distance, I recalled my successful pursuit of a Broad-billed Sandpiper when I was nineteen. I was interning in an oncology lab at Thomas Jefferson University in Philadelphia, and I called in sick so I could drive to New York City and view the bird, the Eurasian wanderer found at Jamaica Bay National Wildlife Refuge the previous day. Then there was the time I borrowed my parents' car to go to a friend's house overnight—or so they thought. Six hours later, I was looking at a Northern Hawk Owl at the Canadian border. Memorable as those and other chases were, my lapwing encounter felt different.

I'd driven farther and seen rarer birds, I thought, *so why do I feel so much better this time?*

And then it hit me: the bicycle. I'd used it for only a portion of the chase, but that eight-mile round trip required coordination and commitment that a car never had.

That was a lot of fun! Maybe I'll buy a bike and start riding it around Boston, looking for birds.

Given the upsides—reduced transportation costs, no-impact exercise—biking and birding seemed a perfect pairing. My mind raced.

Is bicycle-birding a thing? Are there bicycle-birding clubs? Has anyone done a bicycle Big Year? Because that would be the coolest project ever!

A Big Year is an elaborate avian scavenger hunt that runs from January 1st to December 31st and motivates a birder to maximize the number of species observed within a chosen geography. Those constrained by work, family, or finances might undertake a city or county Big Year; those with more free time and money to invest in the adventure might operate at the state, country, or continental level; a privileged few have tackled the entire world. Regardless of the scale and investment, a Big Year is a fun excuse to visit new places and see lots of birds. It's whatever each birder makes it: a weekend distraction, a year-long purpose, or a life-changing journey.

Among many Big Year variations, the "ABA Area"—the lower 48 states, all of Canada, and all of Alaska, as defined by the American Birding Association (ABA)—has emerged as the benchmark.* Multiple birders undertake ABA Big Years each year, and these annual campaigns have become incredibly competitive. Though recognition in the birding community is the only prize for whoever observes the most species, some competitors fly up to 200,000 miles and drive another 50,000. The most invested might chase a Caribbean vagrant in Florida one day and a Siberian vagrant in Alaska the next. An ABA Big Year is a fun undertaking irrespective of budget, but the participant with the biggest bank account has a huge advantage. When the cost of last-minute plane tickets is heaped onto that of rental cars, boat trips, hotel rooms, restaurants, professional guides, and organized tours, some Big Year birders spend well over $100,000 on their campaigns. Top finishers were tallying upward of 750 species in the 2012 moment when the idea of the bicycle permutation first struck me.†

As the ferry churned across Nantucket Sound, I couldn't escape the thought. It was naive and arrogant to think that I, someone with

---

* Hawaii is part of Oceania, and its avifauna is derived from the South Pacific rather than the Americas. It was therefore excluded from the ABA Area until 2016, when political and conservation considerations suggested reevaluation. This narrative unfolds before that inclusion, so Hawaii will be ignored as per convention at the time.

† Going forward, "Big Year" will imply the ABA variety unless otherwise qualified.

zero cycling experience beyond the day's lapwing pursuit, could survive a two-wheeled transcontinental odyssey, but the pull of the birds and the romance of the open road were overwhelming. Contemplating the prospect further, I realized the bicycle would be an egalitarian and environmentally sustainable twist on the existing Big Year model; money wouldn't turn the pedals for me, and my only carbon emissions would be metabolic. Petroleum combustion is a major driver of global warming, so there wouldn't be an inherent conflict between my travels and the birds I'd seek, many of which are threatened by climate change.

Equity and environmentalism aside, I believed a bicycle Big Year would be an amazing adventure. I'd read Kerouac's *On the Road*, Krakauer's *Into Thin Air*, and Kaufman's *Kingbird Highway*, the last a relatable tale of the author's 1973 hitchhiking Big Year, and I felt a sudden urge to become a similar protagonist. The task would be herculean, but I knew the journey would present an unparalleled opportunity for perspective and personal growth, the undertaking a chance to redeem the adolescent birding dreams that alcoholism and academia had crushed across two intervening decades.

My imagination persisted during the crossing but flopped when we docked and the steel gangplank slammed down on the concrete loading ramp. That clatter signaled a return to routine and responsibility.

What the hell was I thinking?

Entrenched in my scientific career, I couldn't convert my biking Big Year dream to reality without obliterating my professional prospects. If I left my research track, then everyone would think I was a failure.

And what if I was injured or killed?

A bicycle Big Year would impact Sonia, my girlfriend of four years, and I wasn't keen to potentially jeopardize our future together. Even if she gave the journey her blessing, my risk would be hers; subjecting her to constant worry and potentially dire consequences if I was struck by a car would be unfair and selfish.

Great. It's settled. I can let this ridiculous idea go.

Except I couldn't. The idea monopolized my thoughts as I drove back to Boston. I parked the car, lugged my gear up the stairs, and pushed through the door to our apartment.

Sonia was snuggled under a blanket on the couch, watching television. "Please tell me you found the birds," she said in a pensive voice.

"Yep! It was an awesome day," I replied.

"Yay! Glad we'll be celebrating instead of sulking." Motioning toward the kitchen, she continued, "I bought ice cream for either outcome!"

While she scooped, I recapped my lapwing coup. Sonia was more nature enthusiast than die-hard birder, but she listened attentively, smiling and engaging throughout. It was wonderful to know I'd found a partner who was as invested in my happiness as I was.

Settling into *Law and Order* reruns with Sonia beside me, I opened my laptop and explored the connection between biking, birding, and Big Years. Google searches revealed Malkolm Boothroyd, a Canadian teenager who, alongside his parents, biked and birded his way through Canada and the lower 48 states in 2007 and 2008. His effort wasn't an official Big Year because it spanned two calendar years, but he observed 548 species along his family's 12,000-mile arc. I also read about Jim Royer, a Californian who'd undertaken a formal January-to-December bicycle Big Year in San Luis Obispo County in 2010, that effort netting him 318 species. But as far as I could discern, no one had attempted a bicycle ABA Big Year by the time the idea dawned on me.

Too excited to contain myself, I unleashed my vision on Sonia. We'd recently rewatched *The Big Year*, a 2011 movie based on Mark Obmascik's book of the same title, so the premise computed immediately.

"Wow! That's a really cool idea! Biking would be way more interesting than using planes and cars like everyone else."

I was spitballing more than selling, so I was surprised by her enthusiasm. "Really?" I replied. "I thought you'd think I was crazy."

"Babe, that ship has sailed," she said, "but you're not ready to trash your career right now, are you? That's what it'll take to make a 2013 Big Year happen, right?"

She could do the math as well as I. Pulling off a Big Year would require me to leave my research position before January, which we both knew I wouldn't be able to do. I'd need months to finish ongoing experiments, catalog reagents, write up findings, and hand off projects, and I wasn't keen to make a hasty, incendiary exit in case the bike trip bombed and I wanted to return to the scientific arena. Likewise, researching a route, managing the logistics, and training would take time.

"Yeah, I know. It's just an idea," I said. "I figured I'd put it out there since I can't get it out of my head."

She replied, "No one says you need to. Tuck it away for another time. The birds will be there if you're ever ready to take the leap."

As usual, she was right. I wasn't ready to make a life-altering decision, and the end-of-year timing offered the logistical cover to default to the familiar path without feeling spineless. It was, ironically, those calendar considerations that absolved us of discussing how a year-long separation could stress or even potentially destroy our relationship. There were more immediate obstacles in place, so that difficult discussion could wait. Until circumstances changed, Sonia, science, and weekend birding around Boston would be enough.

TWO

The Most Important Experiment

S leep had proven elusive for weeks, but that night's tossing and turning were extreme, anxiety antagonizing me as I tried to think about anything but my scientific future. When an electronic throb offered reprieve at five thirty a.m., I donned the same wrinkled shirt and tattered pants as the previous two days, wolfed down a bowl of cereal, and shuffled toward the Red Line station in Davis Square. Sliding doors beckoned me aboard, and I assumed a seat at the vacant end of the car, my tired eyes fixed on the floor as the train departed.

A postdoctoral fellow in the Department of Molecular Biology at Massachusetts General Hospital, I'd spent the previous two-plus years deconstructing synapses, microscopic structures through which neurons communicate. My goal was to identify genes that regulate synapse remodeling, and I hoped my findings would inform research into autism and schizophrenia, cognitive disorders thought to result from remodeling defects. Technically and ethically unable to perform my experiments in *Homo sapiens*, I turned to *Caenorhabditis elegans*, a microscopic worm, as a genetic stand-in; its minimal nervous system facilitated structural analysis, and insights I gleaned in the model system were likely to translate to humans because molecular function is conserved across the evolutionary spectrum. With fascinating biology, important health implications, and powerful technology, the project held huge promise when I initiated it in January of 2011.

Dr. Joshua Kaplan, a world-renowned neurobiologist with a parallel academic appointment at Harvard Medical School, was my advisor at Mass General. Josh had a stellar history of training postdocs and placing them into faculty positions, and he shared my optimism that I could leverage a hypothetical remodeling breakthrough into a tenure-track appointment at a reputable research university. I loved performing experiments, advancing our understanding of biology, and sharing my scientific curiosity, and I'd envisioned becoming a professor throughout my four undergraduate years at Stanford, three predoctoral years at Harvard, and six graduate years at New York University. If everything went well on the experimental front at Mass General, then I'd spend five to six years there before starting my own group at another institution.

Unfortunately, the remodeling problem proved more stubborn than either Josh or I imagined. Technical difficulties guaranteed I spent my time troubleshooting, and unanticipated genetic complexity prevented me from obtaining clean and congruous results. Obsessed with experimental design and conduct, I poured myself into the project, my nerves fraying, work hours swelling, and weekends evaporating as I clawed for traction. By the time Sandy barreled up the East Coast in late 2012, I had only frustration to show for nearly two years of work.

Despite those difficulties, Josh and I generated additional hypotheses post-superstorm. I'd be in business if one of them hit, but the first set of potentials yielded nothing. Fears of postdoctoral purgatory gripped me as I nibbled my fingernails to bloody nubs, and I failed to give my relationship the attention it deserved as my scientific focus constricted and the New England winter deepened.

Seated next to me on our sofa in February 2013, gazing into my sagging eyes, Sonia expressed her concerns. "You're killing yourself for an idealized future. It isn't healthy."

"I know," I replied, "but I gotta work through this. It's my only path to success."

She countered, "But what's success worth if it makes you miserable? And it's affecting me because you're so negative all the time. Minus the time we're hiking or birding, it's like I'm dating Eeyore."

I apologized for my attitude and explained that I was working on a final round of salvage experiments. I summarized, "If something pops, then great; if not, then I'm fucked."

As the train pulled into Charles-MGH station at six a.m., it seemed like Earth's gravity had quadrupled, my shoulders slumping and my feet heavy as I stepped off the train and crossed Cambridge Street. It was late March of 2013, and I'd all but exhausted my final pool of testable hypotheses. Unless the morning's experiments yielded insight, I'd fold the last two-plus years and begin anew with a different project, a surrender that would lengthen my fellowship to at least eight years and greatly diminish my prospects of securing a faculty position.

I entered my lab, flipped on the lights, and navigated humming machines to reach my bench. Shelves of glass bottles towering above me, I readied my first sample, entered a darkened vault, and placed the slide on the microscope, an attractive green glow filling the room when I flicked on the laser. I placed my exhausted eyes on the oculars and used the attached camera and computer to characterize and quantify the worms' synapses. The tedious process repeated for eleven additional samples across the next four hours; I begged the data to reveal something interesting.

When I combined the morning's measurements with those from previous iterations and subjected the pooled data to statistical analysis, I discovered that I'd discovered nothing. Though demoralizing, it was a defeat I'd been anticipating for months. More concerning was the realization that I'd refuted one of the key findings on which my project was based, that genetic condition included in my panel as an extra control. Alone in the dark, staring at numbers on the computer screen, I started hyperventilating.

What the hell is going on? How can I explain the conflict?

Elbows on the desktop, fists tugging at unkempt hair, I rocked back and forth in my chair as I struggled to reconcile the disconnect. I'd performed experiments to replicate the key finding when I joined Josh's lab—taking up the project without confirming that published and foundational result would have been naive—but I'd since spent thousands of hours perfecting the relevant assay. Confident my optimized protocol had revealed true biology on this latest iteration, I felt my face chill as blood drained from it.

Shit! Shit! Shit! This cannot be happening!

If the older, apparently flawed assay implicated the specified gene in the remodeling process—and I'd based all of my hypotheses on that incorrect conclusion—then my downstream experiments would not be informative.

Goddamn it! Now I know why I haven't found shit for two years! I've been looking in the wrong genetic direction the entire time!

I was devastated. My remodeling project died at that moment. Tears streaming down my cheeks, I heaped dirt onto my lifeless postdoctoral child.

The next week was miserable as I collected myself and considered alternatives. A lateral move into biotechnology would double my salary and halve my stress; alternatively, I could slide into teaching or consulting if I was done with experimentation. Despite those and other possibilities, I couldn't bring myself to turn away from the research trajectory. I had to achieve my professorial ambition to avoid being perceived as a failure, so I reluctantly readied to reconceive my fellowship around another project. Too insecure, too scared, and too apathetic to fully reinvent myself, I allowed ego and expectation to force me forward even though my experimental spirit had been crushed.

Watching hockey at a local pub with Sonia the following week, I outlined my retooled postdoctoral plans. "I have a couple of ideas I'm going to pitch to Josh next week," I explained. "None are as exciting as the remodeling project, but I gotta get some positive data flowing."

"What does that mean for the next few years?" she asked.

"More of the same, hopefully with better results."

"But you could be right back here, with nothing, in another three or four years?"

My pessimistic streak had reached new levels, so I replied, "More than likely."

She absorbed that statement without visible emotion. She was like a poker player, silently sizing me up while contemplating her next move. She went all in.

"I want you to get out of there," she said.

I knew she was frustrated with my long hours and weekend work, but she hadn't directly inserted herself until that moment.

"And do what?" I asked with a forced laugh intended to diffuse sudden tension.

"The bicycle Big Year."

I hadn't anticipated that response. I'd spoken casually about the idea since my lapwing triumph in November but hadn't advanced a formalized plan to pursue the imagination. I replied after a stunned pause, "Seriously? You think I should trash my career to ride a bike and look for birds?"

"Don't you see? It'll be the most important experiment you ever perform because it's your best chance to find yourself and be happy. You need it. And you need it now. And I don't want to be the reason that you don't do it."

Through sudden hiccups, I replied, "I don't know where to start. The trip. The lab. Us. I don't know if I could handle a year on a bike, a year away from you."

"I don't know either, but you need to trust and try. The idea of you biking around the country scares the hell out of me, but we have to face the journey together if it's going to make you the man I know you can be. I'll be here when you get back," she reassured me. "This has been one of the hardest decisions of my life, but you need to know the door is open if you want to walk through it."

She'd saved my life once before and was trying to do it again. Tears welling, I whispered weakly, "I love you so much. I'm going to miss you more than you'll ever know."

"Just do everything you can to come back to me. I don't know what I'd do without you."

Two weeks later and just days after the Boston Marathon bombing on April 15th, that event a sobering reminder that tomorrow is never guaranteed, I walked into Josh's office and issued my declaration. "I've had enough. I'm done," I said against swelling emotion.

Josh sat back in his high-backed leather chair and crossed his arms. Singularly committed to understanding the synapse, he preferred perusing papers and digesting data to social interaction. He displayed charm and dry wit when pried from his office sanctuary, but I'd had difficulty connecting with him during my time in his lab, his focus on other individuals and their more fruitful projects.

"I've been anticipating this discussion," he said, "and I think you're making the right decision. You could restart with another project, but we both know you'll be happier doing something else."

I didn't expect that Josh would plead to retain me, my intractable remodeling project dragging on his grants, but I was surprised he'd framed his response in personal rather than scientific or financial terms. I sometimes thought his intense deconstruction of the synapse ignored the higher-order manifestations of the neural connections that he studied, but it was encouraging to know he understood my position.

Reassured, I spoke evenly, "I sensed my project was ruined for some time, but the recent debacle crushed any desire I had to restart with another."

"So, what have you decided? Teaching? Biotech? You'd be great in either. I'm happy to help with placement," he said.

"Thanks, Josh," I replied. "That means a lot, but I'm going to try something different."

Raising an eyebrow, he sought clarification. "And what is that?" he asked.

"Have you ever heard of a Big Year?"

"No."

"Well, lemme tell you about it."

I articulated my intention. Josh's brow furrowed as he tried to comprehend my dramatic departure, but I wasn't concerned if the premise and details computed because calendar considerations and financial concerns were more pressing.

"All that brings me to this," I said, "I need income between now and next January, so I'd like to stay on as experimental support until then."

Appreciating that my technical abilities would accelerate other ongoing projects, Josh agreed. An arrangement struck, I thanked Josh and walked out of his office with a new view of my future.

The remainder of spring was bliss. I enjoyed helping my labmates advance their experimental aims—and thereby their careers—and I relished the additional time I spent with Sonia. I hadn't taken off three consecutive days in eighteen months, so a mid-summer trip to Newfoundland was a welcome escape. We drove from Boston and took in views of puffins, eagles, warblers, moose, and caribou as we camped our way around the island for two weeks.

Returning from Canada, I borrowed a junky bike from a friend and started commuting the seven miles to and from my lab each day. Pedaling along the Charles River bike path, I thought about my upcoming adventure.

As the first person to attempt a bicycle Big Year, I had neither an established route nor a species benchmark to guide my planning. Turning to petroleum-powered Big Years for reference was pointless; flying from New York to Los Angeles is a six-hour formality, but cycling between those points requires an uninterrupted month of 100-mile days. That daunting reality suggested I'd need to reinvent birding around the limits of my legs.

While petroleum-powered Big Year birders experience bouts of birding punctuated by transcontinental travel, the bicycle would require a steadier approach. I couldn't fast-forward or rewind to faraway places with the nonchalant swipe of a credit card, so my route had to be set before I started. Once I committed to a national trajectory and timeline, only regional tweaks would be possible. I'd need to remain flexible against fatigue and injury; simply surviving a bicycle Big Year would be a monumental achievement. Infinitely more people had completed the Appalachian Trail or summited Mount Everest by the time I set to planning my first-of-its-kind adventure.

With those considerations guiding me, I poured through birding guides and resources in the months preceding my January jump-off. I wouldn't be able to chase individual vagrants across the continent, so I focused on the regularly occurring species, each of which occupies an appreciated geography—its range—based on its evolutionary history and habitat preferences. Ranges are displayed as seasonally colored maps in field guides, migration shifting many distributions through the year, and I cross-referenced hundreds of the graphics to design my route.

I realized I wouldn't have time or energy to pursue every species, so I identified areas where multiple range maps overlapped at each season of the year, my goal to maximize species while minimizing riding.

My first pass suggested I could find about 570 species if I focused my efforts in the climate-comfortable confines of the southern and western United States. That number wasn't exciting, so I revisited the range maps, where a winter visit to New England emerged as an additional but painful possibility. If I braved frigid conditions and found 20 cold-weather specialist birds, then I'd raise my ceiling to 590. Whatever route I rode, I figured I'd intersect ten vagrants by pure chance, so I included New England because it brought the tantalizing possibility of 600 species into play. Weighing a January New England start against a December New England finish, I calculated I could connect the other seasonally constrained geographies with less riding if I confronted the cold at the outset, my Boston home base fortuitously streamlining departure logistics versus starting anywhere else.

From Massachusetts, I'd use the remainder of January and February to ride south to Florida for March before rounding the Gulf Coast in April. Southwestern deserts would be beautiful in May, and the Rocky Mountains would offer incredible riding through June, July, and August. If I survived that high-altitude ass-kicking, then I'd intersect the Oregon coast in September, move south through California in October, and use November to return east and reach the Rio Grande Valley in South Texas. Assuming I had energy left, I could follow the Texas coast north and cut inland, toward Oklahoma, to end the year. Putting it all together, I estimated I'd need to ride at least 15,000 miles to have a shot at 600 species. Staring at a paper map of the lower 48 states, my ideal path traced in blue marker, the task looked impossible.

As 2014 neared, Sonia and I crammed our minimal possessions into a storage locker in the Boston suburbs. Sonia was adventurous in her own right, and she'd planned to travel in my absence. Working remotely as a corporate travel manager, she'd crash with friends and family as she moved around the lower 48 states, our paths crossing whenever proximity and circumstances allowed. Among an endless list of qualities, her confidence and independence were particularly magnetic.

On December 31st, we put my bicycle into the trunk of our car and drove forty miles north to Salisbury, Massachusetts, where two local birders had offered to house us ahead of my departure the following morning. Lying in bed, my arms around Sonia for what I hoped wouldn't be the last time, I asked, "What do you think? Will we make it through this?"

"I think so," she replied. "We've been through so much already, but it's good to keep testing ourselves. If you come back to me, then we'll know this is right, right?"

"Yeah, that's how I feel. I just hope I figure myself out along the way. I mean, I need to be happy for us to be happy, you know?"

"Totally," she said. "And you need to know the trip isn't going to go exactly as planned. Just trust that it'll work out somehow. Now try to get some sleep. The rest of your life starts tomorrow."

THREE

Bicycle Baptism

Staring out the kitchen window and into the predawn darkness, I recoiled when I glimpsed the outdoor thermometer.

Twenty-nine degrees? Sure. Nineteen degrees? Maybe. But nine degrees? Kill me now.

Incredulous, I shuffled toward the back door and stuck my bedhead out. Mother Nature satisfied my curiosity with an icy, backhanded slap across my face, and I slammed the portal shut before she could extend her sub-freezing shellacking.

Nine degrees it is. Damn.

I knew starting my bicycle Big Year in Massachusetts could expose me to crippling cold, but mild temperatures through November and much of December had sowed hope of a balmy beginning. That optimism suddenly frozen solid, I gobbled a bowl of oatmeal and prepared to confront my meteorological misfortune.

I layered long underwear, fleece thermals, and canvas cargo pants before donning an insulated undershirt, a fleece jacket, and an expedition-weight down parka. I protected my hands with mountaineering gloves and stuffed my feet into thick wool socks and stout winter boots. With a neoprene face mask leaving only my eyes exposed, I looked ready for a polar bank robbery, my bicycle getaway to be the slowest in heist history.

In the garage, I made final adjustments to my four red panniers and attached them to metal racks on the front and rear of my bicycle. Functional over flashy, my maroon Surly Disc Trucker was the two-wheeled equivalent of a Hyundai Elantra; featuring basic shifting levers, low-tech derailleurs, and generic parts, the steel-framed touring workhorse would be easy to maintain and repair as I moved around the country. I bent at the knees, gripped the frame, and hoisted my loaded transport. Accounting for tools, binoculars, camera, spotting scope, tripod, bird book, clothes, food and water, first aid kit, toiletries, laptop, chargers, and other supplies, it weighed nearly eighty pounds. I hadn't pedaled an inch, but my legs were already aching.

My cycling experience at the time I left Salisbury was laughable. I'd biked the fifteen round-trip miles from my apartment to Mass General through the summer and fall of 2013, but that minimal training was halted in October, when I was stricken by tendonitis in my knees. Unsure if I'd rehabilitate before January, I delayed purchasing my bike and gear until early December, when physical therapy alleviated my throbbing pain. Holiday hoopla took precedence over riding at the end of the year, and I realized I hadn't practiced riding the fully loaded beast by the time I pushed it out of my lodging on that inaugural morning. I flipped on my bike light and mounted up, the frigid air stabbing at my lungs as I embarked on my Big Year.

The unpaved driveway was an unanticipated first test. My skinny tires sank into the pebbles, and I struggled to control the laden bike as it slipped and slid. I narrowly avoided colliding with the trunk of a large tree in the first twenty feet, and my panicked overcorrection nearly sent me into a low hedge two seconds later. Too stubborn to dismount and push, I wobbled through the gravel until I reached the paved road. My steering steadied, and I slogged toward Salisbury Beach State Park, a coastal reserve four miles east of town.

Purples and pinks overtook the horizon as I neared the beachfront, and I could see rolling dunes to my left and marshes to my right as I turned onto the park's entrance road. I'd visited the area previously, each time by car, but the bicycle offered a heightened sensory experience. Winter wind whispered as it blew through my helmet, the salty smell of the ocean filled my nostrils, and sea spray kissed my numbed lips through the face mask. Every inhale drew the landscape into me, that essence returned to nature as a condensed puff at the end of each labored breathing cycle.

Beyond my frosty, vaporous breath, I glimpsed a ghostly form in the residual darkness. The apparition floated across the road, and I was so entranced that I lost control of my bicycle and drifted off the road before sandy bumps snapped me to attention. I jammed on the brakes short of the reeds, but my rear end slid off the seat and dropped onto the frame during the haphazard deceleration. Disregarding shooting pain in my nether regions, I hurriedly excavated my binoculars from beneath my layers. Focusing the optics as the figment glided into the marsh, my brain confirmed what my heart already knew: Snowy Owl for Big Year bird #1.

Twenty-plus inches long and boasting a wingspan approaching five feet, the Snowy Owl is one of the most badass birds on Earth. Widespread on the Arctic tundra where they breed, the diurnal sight-hunters subsist primarily on lemmings, their crushing feet and needle-sharp talons terrorizing rabbits, waterfowl, and fish, beyond their primary rodent prey. Their broad wings propel their hulking bodies across the landscape with an artful buoyancy, and their piercing yellow eyes infiltrate an observer's soul with a single stare. They steal breaths and stop hearts, and every birder remembers their first view of the incomparable killer angel, mine on a snow-covered beach in New Jersey when I was fourteen.

Accessible only to adventurous and financially able birders during their Arctic summers, Snowy Owls migrate to central and southern Canada for the nonbreeding months. A few reach the Pacific Northwest, the Upper Midwest, and New England in an average winter, and it's usually possible to intersect with the species by investing a few days at traditional haunts. The wintering birds prefer treeless expanses—farm fields, deserted beaches, runway medians—which mimic their summer tundra digs.

Fortuitously, the winter of 2013–2014 was anything but average as far as Snowy Owls were concerned. The season had unexpectedly unfolded as the largest invasion on record, and by the time of my departure, hundreds of owls had descended on the northeastern United States. They'd appeared in backyards and been spotted on billboards; an extraordinary bird even reached Jacksonville, where it delighted Floridians for weeks. Such winter irruptions are rare and unpredictable, the influxes presumably influenced by food shortages farther north, and birders in the lower 48 invariably welcome the periodic pushes. The historic circumstances suggested I'd find the bird in my first week, but intersecting it as my inaugural species was an auspicious beginning.

My heart pounded as the magnificent specimen lit on a log. I'd seen many Snowy Owls during my three years in Boston, but the species never failed to captivate. Staring at this snow-white specimen, his yellow eyes returning my gaze, I could almost believe I'd been transported to the Arctic. That's the magic of birds; they inspire connection to the natural world while motivating us to imagine the planet from their perspective.

I begrudgingly left the Snowy and continued to the park's southern end, where two jetties flanked the outflow of the Merrimack River. Like extended beckoning arms, the rocky structures calmed the roiling Atlantic and welcomed wintering birds into their protective interspace. Assembling my spotting scope and scanning the inlet, I spied a raft of snow-white Common Eiders bobbing playfully on the waves and a lone jet-black Great Cormorant standing a statuesque vigil on the jetty's terminus. Loons, grebes, and mergansers fished in the boating channel, and gulls floated along the beachfront. It was wondrous that birdlife abounded in such harsh conditions.

Dutifully examining the gulls, I noticed an outlier. Its size, plumage, and behavior suggested a Black-headed Gull, a Eurasian species that occasionally wanders across the North Atlantic; better views as it approached the jetty confirmed that exciting suspicion. I nearly slipped into the sea during my subsequent victory dance, but I steadied myself on my tripod and avoided plunging headlong into the surf and the accompanying hypothermia. I'd planned to follow up on reports of the species as I moved south through New

England and the mid-Atlantic, so I was stoked to secure the bonus bird without pedaling miles out of the way. I'd not projected the foreign visitor in my total, so it was an unexpected bump toward 600 on my first morning.

I spent an additional hour at the inlet before remounting the bicycle and departing the reserve. The world was awake by midday, and cars whizzed by me as I labored north on Route 1A, renewed blood flow returning feeling to my hands and feet by the time I crossed into New Hampshire. Birding the seasonally vacant beach communities en route to Hampton, I found several of the high-latitude species that motivated my New England start: Snow Bunting, Lapland Longspur, Glaucous Gull, and Purple Sandpiper. Like the Snowy Owl, those Arctic nesters reach only as far south as the mid-Atlantic in winter. The faster I found those and other cold-weather targets, the sooner I'd be able to head south, toward warmer climes.

By the time I returned to my lodging in Salisbury, Massachusetts, that evening, I'd found thirty-nine species, familiarized myself with the loaded bicycle, and logged twenty-five miles in temperatures peaking at twenty-seven degrees. Mine was a bone-chilling baptism, but red-hot birding kept me motivated from start to finish. A steamy shower concluded my initiation, and I enjoyed a delicious dinner with my fifty-something hosts, Henry and Deb.

Stomach full and eyelids heavy, I retired to the guest room and recounted the day's events—complete with bird photos and a map of my movements—on my blog. I'd christened *Biking for Birds* several weeks prior to departing and used a series of introductory entries to describe the project and its goals. My hope was to build the personal journal into a networking platform as I progressed, but word of my impending adventure spread through the birding community faster than I'd imagined it could. By January 1st, I'd received hundreds of well-wishes and dozens of housing offers around the country. Given the challenges I'd face, it was reassuring to know I'd have support in many of the areas I planned to visit.

My daily chronicle complete, I climbed into bed and flipped out the light. The day had unfolded perfectly. Apprehension yielded to excitement, and I was eagerly anticipating the second day of my adventure as I nodded off to end the first. By the time I awoke eight hours later, it was a completely different story, my optimism crushed in a single Arctic blow.

FOUR

The Polar Vortex

Just after midnight on January 2nd, a gargantuan Arctic air mass descended on the northeastern United States. Branded a polar vortex by meteorologists, the historic low-pressure system was forecasted to plunge New England into sub-zero temperatures and bury eastern Massachusetts in snow, three inches of which had accumulated by the time I awoke. I made my way to the kitchen, where my hosts were sipping coffee.

"Sleep OK?" asked Henry.

"Like a rock. Didn't realize how tired I was after yesterday," I replied.

"Well, settle in," Deb said, "there's more snow coming."

"And don't worry about staying another night," Henry added. "You're welcome indefinitely."

I'd planned to spend the day at Plum Island before overnighting with another birder in nearby Newburyport, but the weather crushed that intention. My ambition caged, I paced the living room while picking at my cuticles, the whiteout intensifying through the afternoon and evening. By the time I bedded down, my Big Year dream was buckling under a foot of flakes and an avalanche of anxiety.

I rose early on the 3rd, nervously pulled back the curtains, and absorbed a white world, the garden engulfed and the driveway disappeared. And it was still snowing.

OMG! I'm going to be stuck here forever!

I joined Henry and Deb for breakfast. "What's the latest word on the weather?" I asked.

Deb replied, "It's supposed to let up around noon, so hopefully that'll give the plows time to clear the roads before tomorrow."

"Good news is the bird feeders should be really busy this afternoon," Henry said.

He was right, that burst of activity a needed distraction as I observed it from the living room sofa. When seed levels dwindled midafternoon, I bundled up and pushed through thigh-high drifts of snow to refill the buffet. Stomping around the backyard, dispensing thistle and millet into the various tubes and platforms, I felt like I was reliving my childhood.

———

My birding interest sprouted at age seven, when my family moved from the concrete confines of downtown Philadelphia to the more suburban, Chestnut Hill sector of the city. Our half-acre lot felt like untamed wilds compared to the puny brick patio at our former address, and I was immediately struck by the array of feathered forms populating our property, many of which visited our backyard feeders.

Armed with an old pair of binoculars I excavated from Dad's possessions and a bird field guide Mom purchased at a yard sale, I confronted the backyard frontier. There was something new and beautiful in every direction I turned, and I slowly learned how to differentiate sparrows, finches, wrens, and woodpeckers. A portly White-breasted Nuthatch made me laugh as it walked head first down the trunk of a pine, and a luminous Northern Cardinal demanded attention as he sang from a rhododendron. Wonderful as those were, none trumped the charismatic Blue Jay; with a black necklace offsetting strokes of cobalt, azure, and cerulean, the bird was the stuff of dreams, my rudimentary sketches a poor representation of evolution's artistry.

To my parents' delight, birding kept me occupied and out of the house. Less to their liking was my parallel pastime of hurling rocks, crabapples, and snowballs at the commuter trains that ran directly behind our property. It was, I think, that mischievous streak that spurred my "indoorsy" guardians to support my outdoor interest; had they not supplied upgraded binoculars and endless bird books, my boyhood balance might have tipped toward railway vandalism, a legitimate concern once I discovered the unadulterated joy of dropping Halloween pumpkins onto the tops of trains from a nearby footbridge. Between the birds and the railway, our neighborhood was heaven for a curious and mischievous boy.

By the time I was ten, my parents were willing to deposit me at natural areas for hours at a stretch while they ran errands or attended to social calls. I had no cell phone, and their willingness to let me roam the world unsupervised contrasted with the "helicopter" approach other parents employed. Those early forays expanded my birding foundation, fomented a deep appreciation of the natural world, and sowed unusual preadolescent independence.

My passion grew to include local nature centers, the Jersey Shore on family vacations, and a series of summer birding programs, the first in Arizona when I was twelve years old. Camp Chiricahua was my first foray into the American West, and my head nearly exploded when I viewed Greater Roadrunners, Elf Owls, Elegant Trogons, and Red-faced Warblers for the first time. Wonderful as the birds were, the people were the best part of the experience. I'd known only one other young birder before Camp Chiricahua, so the collective energy and enthusiasm of the other dozen campers was infectious. It was from those others, while we were piled into the van and commuting between birding locations, that I learned about Big Years. I formed strong bonds with our leaders, fifty-something Victor Emanuel and nineteen-year-old Barry Lyon, and the two weeks I spent at Camp Chiricahua were the happiest of my life to that point.

My avian fascination strengthened through middle school, while classes in physical and life sciences familiarized me with the scientific method and offered an evolutionary framework into which I placed my swelling obsession. I was introduced to ornithologist Robert Ridgely from the Academy of Natural Sciences in Philadelphia—Mom was in the same gardening group as his wife—and the stories he told about discovering and conserving birds in South America suggested I could employ my compulsion to professional ends. By age fifteen, I was resolute on blending birding and biology into a vocation. I also thought a Big Year would be fun, and I envisioned participating in the transcontinental scavenger hunt at some point, perhaps after I graduated college.

I returned to the living room after refilling the feeders. Sipping hot chocolate in front of a roaring fire later that evening, I couldn't believe the detour I'd taken between those initial Big Year musings and my bicycle undertaking. Alcoholism and academia had delayed those adolescent dreams nearly two decades, and I stopped worrying about the weather when I realized that twenty inches of snow presented a small challenge when compared to what I'd already overcome. Reassured by bedtime, I flipped out the light and drifted off.

I couldn't find Henry and Deb when I woke, but I saw a sport utility vehicle turn into the driveway while I was foraging in the kitchen. I met them at the front door and helped unpack bags of food to replace what we'd consumed during our confinement.

"You wouldn't know it from looking out the window, but Route 1A is plowed," Henry remarked.

The street in front of the house hadn't been cleared, so I assumed I'd be housebound through at least the remainder of that day, January 4th. I sought clarification. "No joke? Is it clear enough to ride?"

"Maybe," Deb said. "The shoulder is mostly buried, but you could use the outer edge of the traffic lane. Not many cars on the road, so that's good."

"Give it a try. Just come back here for another night if it's a disaster," Henry suggested.

With nothing to lose, I scarfed down my breakfast and prepared to test the roads. The outdoor thermometer indicated an excruciating minus-ten degrees, but I felt pressure to ride after two lost days. I did add 14 new species on the backyard feeder during the delay, pushing my total to 53, but returns would diminish as I exhausted the common birds in coming days.

Bundled beyond my New Year's debut, I thanked my hosts for their extended hospitality and staggered out the door. An Arctic blast infiltrated my balaclava, and my face absorbed a million icy pinpricks as I pushed my bicycle down the driveway and into the road. The cold stiffened my joints, and my feet searched for purchase as I forced the stubborn contraption through knee-deep snow. I felt like I'd been thrust into a reality television physical challenge, my string of uncoordinated flops and headlong dives suggesting my imminent elimination, but I made labored progress toward Route 1A. It was clearer than I anticipated, so I decided to give it a go.

Crossing over the frozen Merrimack River, I pedaled south through Newburyport, a quaint colonial seaport with seventeenth-century roots. Brick storefronts and wooden homes lined my route, and road conditions held as I powered out of town and into more rural surroundings. Industrial-scale farming has bypassed New England, and stands of leafless woodlands and snow-covered stone walls lent definition to the individual homesteads that comprised the agricultural patchwork. With wispy smoke rising from the chimneys of snow-caked farmhouses, I felt like I was cycling through a Robert Frost poem, middle school echoes of "The Road Not Taken" urging me forward.

When the breeze shook snow from overhead pine boughs, I felt like I was sealed inside a gigantic snow globe. Peering beyond the drifting flakes and through my imagined encapsulation, I saw my fifteen-year-old self staring into the curio as he tried to understand the path I had chosen. His future was my past, but an understanding of the time separating us was still many months and thousands of miles away. Trusting that the road would reveal more on its own timetable, I veered southeast on Route 133 and rolled onto Cape Ann midafternoon.

Located northeast of Boston, Cape Ann boasts a rugged coastline dotted with rocky promontories. The two main towns on the peninsula, Rockport and Gloucester, differ in history and character. Rockport is smaller and more charming; a supplier of timber and granite in the nineteenth century, the town evolved into an artists' colony and summer retreat. On the southern side of the peninsula is Gloucester, a larger, grittier, and historically fishing town that has suffered from the depletion of the cod stocks in recent decades. Gloucester has its own attraction, but its elevated opioid abuse rates suggest the city struggles in ways Rockport does not. Both towns are quiet in winter, their frigid surroundings offering some of the continent's best cold-weather birding.

Among the birds that frequent Cape Ann in winter are alcids—the family that includes auks, puffins, auklets, guillemots, murres, and murrelets. Restricted to the northern hemisphere, alcids occupy the same ecological niche as southern hemisphere penguins; they share black-and-white plumage schemes, strong swimming abilities, and endearing terrestrial ungainliness. Alcids, however, can fly, and several species migrate from their Canadian breeding grounds to New England waters each winter. Staring out to sea through a spotting scope is the best way to find alcids, and I secured views of three species—Razorbill, Black Guillemot, and Dovekie—without much effort. My run at a fourth, the Thick-billed Murre, was more dramatic.

In the lead-up to my adventure, I disseminated my phone number to local birders so that they could call me if they found anything unusual while I was nearby, each day's route outlined on my blog the previous night. Intrigued when my phone rang on the morning of Sunday the 5th, I removed my gloves, fished the device out of my coat pocket, and greeted the unidentified caller, "Hello?"

A timid, adolescent voice inquired, "Is this Dorian?"

"Yep. What's going on?"

He introduced himself as Miles Brengle, a teenage birder whose name I immediately recognized. He continued, "A Thick-billed Murre just swam into Gloucester Harbor. It's visible from the wharf."

Thick-billed Murres winter in Massachusetts, but they're usually so far from shore that a motorboat is required to observe them. It was a species I thought I might find inshore, but only with timely assistance from locals.

"Hell yeah! I'm on my way," I exclaimed. "Keep an eye on it, kid!"

I abandoned Rockport and began the six-mile sprint to Gloucester. Fearing the murre would be flushed by a passing boat, I rode with urgency, caution compromised as I streaked over patches of ice, around head-high snowdrifts, and straight past stop signs. I almost flattened several pedestrians, and I deftly avoided a large truck when it reversed without warning. Frozen snot streaking across my numbed face, I extended my chase on aching legs.

I was on the verge of collapse when I reached the wharf. I dismounted and stumbled toward a group of six birders, Miles among them. Bending down and peering through his spotting scope, I breathed a huge sigh of relief when I saw the murre, the snoozing bird oblivious to the surrounding harbor clamor.

I turned toward Miles. "Too bad I'm an alcoholic and you're only fourteen. Otherwise, we'd crack open a beer to celebrate!" I said.

Miles smiled sheepishly while the adults chuckled. All the faces were familiar, and the ensuing twenty minutes evaporated as the assembly inquired about my nascent adventure. Though an insignificant fraction of birders bow to competitive impulses by keeping information about rare birds to themselves, the overwhelming rest are friendly and cooperative. Sharing sightings strengthens the communal experience, and I was happy the murre had allowed Miles to join my story. Watching him school the adults on the various harbor birds, I saw much of my adolescent self in him. Our interaction was a welcome reminder of the ageless connections that birding facilitates.

I eventually excused myself and cycled three blocks to a Dunkin' Donuts. I leaned my bike against the storefront and entered under the watchful eyes of what looked like eight local fishermen, the motley crew spread across several tables and discussing the Patriots' playoff chances. I paid for my sugary snacks and claimed a table. A huge sports fan, I'd hoped to eavesdrop on their playoff prognostications, but they clammed up as soon as I sat down.

A guy with hands like leather and teeth like a picket fence addressed me with a Boston accent. "Hey you, Mr. Stay Puft, whaddya doin' on that bike?" he asked.

I answered while chewing on a Boston cream. "I'm riding 'round the country and looking for birds. Just started five days ago. Cold as fuck out there."

A second man followed up from under the tattered brim of a salt-stained Red Sox hat. "So, you're like those crazies on the wharf in the middle of winter?" he asked.

"Yeah. But crazier. I'm going to try to go everywhere on my bicycle. Here, Florida, Texas, Colorado, even California."

At that point, I may as well have had two heads; their glassy eyes and blank stares suggested me a sideshow freak or—worse—a New York Yankees fan. I searched for another language to connect with them. "How many of you guys fish?" I asked.

Several offered reluctant but affirming hand gestures. Others conceded defeatedly, "Use ta."

"Well, birding is a lot like fishing," I said. "I used to fish a lot as a kid, so I know what it's like to put out a line and hope something bites. While you're waiting, you dream a bit, and whether or not you catch a fish suddenly doesn't matter as much. That's what birding is like for me. The bike is my boat, and the road is my ocean. I never know what I'm going to find, and that keeps me going."

Heads nodded, and I caught a crooked smile peeking out from beneath the Red Sox hat. The men invited me to join them, and I fielded a string of questions. They wanted to know what I did for work before I left, how I planned to survive a year on a bike, and what I would do when my journey was over. I answered what I could and told them to check back about the rest. As I polished off my second donut, the guy in the hat shouted to the woman behind the counter, "Hey, Sandy! Get this guy another doughnut on me! Crazy son-of-a-bitch is going to need it!"

The prospect of adventure and the unrealized degree of personal freedom intrigued the men more than the birds, and their curious tones and envious eyes betrayed the fact that they were likely locked into their current lives for familial and financial reasons. I respected them for honoring those obligations but was silently thankful for the freedom my academic departure and Sonia's emotional support afforded me. For one year, my imagination and the strength in my legs would be my only governors.

Birding eventually beckoned, and I bid my new friends goodbye. They wished me safe travels and returned me to the world with a glazed chocolate doughnut snug in my pocket. With a crunchy shell surrounding a sweeter and softer inside, it was exactly like the providers.

I'd earmarked that afternoon to search for a King Eider, an Arctic duck that reaches the lower forty-eight states in small numbers each winter. Though the female displays muted brown plumage, the male boasts a black and white body offset by a powder blue crown, a pastel green cheek, a bulbous orange forehead, and a bright red beak; I couldn't conjure such a colorful creature even with the aid of hallucinogenic mushrooms. It's rare to intersect the species outside the Arctic, but a male, presumably the same one, had been known to winter on the eastern side of Gloucester for seven consecutive years. He was the most predictable King Eider on the continent, and I knew exactly where to find him because I'd seen him many times in the previous two winters.

I cycled to a low seawall and set up my spotting scope. A stiff breeze and choppy seas made viewing difficult, but I eventually spied the bird bobbing a half-mile offshore. Even from that distance his head was stunning, a psychedelic splash of color in a seething blue abyss.

More than his appearance, he engendered a great sense of wonder. I had no idea where among Alaska, the Canadian Archipelago, or Greenland he summered, but the fidelity with which he returned to Gloucester was remarkable. Migration is fascinating, and research suggests that the behavior is genetically hardwired. That discovery doesn't diminish the intrigue, but it did suggest that the eider's biannual translocation was an unconscious outcome.

Marveling at the rainbow King, I thought about how different humans are from birds. Our higher-order cognitive abilities subject our behaviors to conscious control, and a complex interplay of emotion and logic influences our actions. We possess the capacity to examine our purpose and evaluate our progress, and our ability to make calculated decisions toward desired ends is our most uniquely evolved trait.

Staring out to sea, my teeth chattering like an old-fashioned typewriter, I was happy I'd found the strength to trade pipettes for pedals. I'd grown tired of unconsciously commuting between my apartment and my laboratory, but fear of failure had prevented me from looking beyond familiar but stifling confines. There were more traditional exits than a bicycle Big Year, but I knew the journey would be a powerful microscope to examine myself at a confused time.

I didn't know what to expect when I pedaled into the Salisbury darkness five days ago, and no amount of planning could have prepared me for the diabolical polar vortex. In conditions that should have thwarted me, I'd learned to handle the loaded bicycle and found many of the cold-weather birds that motivated my Massachusetts start, the remainder to be sought as I moved south through New England and the mid-Atlantic in the ensuing weeks. The bicycle offered a constantly changing view of the world; roadside interactions lent perspective the laboratory had denied. My decision to leave academia felt right, and I couldn't wait to see how the upcoming weeks would unfold. Departing the eider and the routine he represented, I turned my attention south, the 2,000-mile ride to Florida feeling more opportunity than obstacle as I wheeled out of Gloucester on the morning of January 6th.

FIVE

Cookies and Conversation

I reached Charlestown, Rhode Island, under heavy snow on the after-noon of January 10th. I pulled into the driveway, leaned my bike against the side of the one-story colonial house, and rang the doorbell. I knew little about my host, Bobby, save that he was a local birder who contacted me after I posted a link to my blog in an online discussion forum. His email reply indicated he was curious to learn more about my adventure and communicated that he and his partner, Susan, would be happy to host me as I commuted south.

The door opened, and a diminutive woman with close-cropped white hair was revealed. Appearing to be in her eighties, she grabbed my arm, whisked me into the house, and closed the door behind me.

Dusting snowflakes from my body, I greeted her, "Thanks, Susan."

She snickered and replied, "That's a mistake a lot of people make. I'm Bobby."

Realizing I'd only communicated with her via email, I reoriented gender expectations just as Susan poked her white-haired head into the foyer. She was equally friendly, and the two women shepherded me to the guest room, where they left me to clean up ahead of a highly touted hot supper.

I was beat. The forty-five miles I covered along the convoluted western shore of Narragansett Bay represented a new daily high for distance. Temperatures in the teens and twenties limited my enthusiasm throughout the day, and impeding wind and swirling snow rendered the final fifteen miles a cautious crawl. A hot shower redeemed the day's efforts, and I passed out on the bed for the next hour.

I woke in the early evening and joined my hosts at an antique table in the dining room. Bobby and Susan were delightful. Our dinner conversation wound through our shared birding interest and my nascent adventure before settling into politics, specifically the nation's evolving view of same-sex marriage. I'd cheered the demise of DOMA—the Defense of Marriage Act—that unconstitutional legislation defining marriage exclusively as a union between one man and one woman, but my view of homosexuality stretched only as far back as the experiences of my thirty-something cohort. I'd not spent any time with octogenarian lesbians before, so I was fascinated by Bobby and Susan's perspectives as they described what it was like to be lesbian in the 1940s.

"It was a different time. We had to live in silence to avoid criticism and harassment," Bobby explained.

"And there were no organized support networks," Susan lamented. "I had nowhere to turn as I tried to understand my feelings."

Bobby hailed recent progress toward equality, noting, "Policy can change overnight, but ingrained biases take longer to erode."

Taking Bobby's hand, Susan added, "I'm sure we'll get there eventually. I just hope we're still around when it happens."

The two women glanced at one another for a few moments before Bobby turned to me and broke the silence. "You were probably hoping that if you got lucky enough to spend the night with two women that they'd be a lot younger than us, huh?" she said with a sly wink and smile.

Susan dropped her fork onto her plate, threw up her hands, and countered in earnest. "Jesus Christ, Bobby!" She exclaimed, "Spare the poor boy that image! If the cold and snow didn't kill him, that certainly will!"

It took every ounce of self-control to avoid choking, and the women patted me on the back to make sure I didn't die from a fit of laughter. Three of my grandparents passed before I was six, the last—in faraway England—expiring when I was sixteen, so I had very little contact with the elderly through my teens and twenties. I'd sometimes viewed aging as an unwelcome, even painful process, but Bobby and Susan forced a recalibration. They were inspiring previews of seniority, and I hoped time would grant me similar wit, enthusiasm, and grace. Thirty-five years of life behind me, I felt that whatever remained had limitless potential after dining with the two women.

Freezing rain promised hellish riding on the eleventh, so Bobby and Susan insisted that I delay a day with them, an invitation I accepted without hesitation. They devoted much time to ensuring my comfort and offered so much conversation and so many cookies that I wanted to stay forever. When improved conditions permitted departure on the twelfth, they returned me to the snowy world, my heart forever warmer after my time with them.

The latest Arctic assault stung my chapped face, but I withstood icy roads and frigid winds and powered east to New London, Connecticut, by midafternoon. Approaching New Haven two days later, I added Fox Sparrow, a reddish-brown ground-dweller, as my hundredth species. I knew new birds would be harder to find as I progressed, but I devoured a large pizza and a pint of ice cream to celebrate the milestone.

Continuing southeast along the Connecticut coast via Fairfield and Greenwich, I prepared for my looming transit of New York City. I harbored no desire to explore the metropolis by bicycle, but the bike path on the George Washington Bridge at Manhattan's northern reaches was my only legal and petroleum-free way to cross the Hudson River. Knowing I'd have to ride into the heart of the Big Apple, I'd timed my transit to a weekend, when traffic would be less of a threat.

Sadly, gains from that premeditation were negated by another snowstorm, the fourth I'd suffered in eighteen days, and I strayed onto an entrance ramp for the subterranean Cross Bronx Expressway in whiteout conditions. The road narrowed, and I started shitting bricks as traffic bore down on me. I glanced over my left shoulder to evaluate a retreat but retracted my neck like a turtle when I saw a truck mirror coming at my face, my reflexive pull-back sending me into the concrete barrier to my right. My panniers scraped against the containment, and I battled to keep the bicycle upright. I squeezed my brakes, thoughts of Sonia flashing through my head as I brought the bicycle to a careening stop.

Pinned at the margin by a continuous parade of speeding vehicles, I dismounted and used all my strength to lift my loaded bike over the first of two retaining walls. The interspace was a rancid mix of festering trash, broken glass, and human refuse. I was convinced I'd reached my *Law and Order* moment, the drama's iconic "dun-dun" inescapable as I scanned the area for dead bodies. None evident, I dragged my bicycle through the nauseating milieu, hauled it over the second retaining wall, and scampered off the adjacent ramp ahead of speeding traffic.

My heart racing, I exhaled when I reached the refuge of a nearby sidewalk. A navigational blunder had nearly cost me my life, and I felt unexpectedly fragile as I propped my bicycle against the side of a building and my adrenaline slowly equilibrated. I'd long assumed any accident I might suffer would be the fault of a careless driver, so the event was a harrowing reminder of my responsibility. Though I'd promised Sonia I'd stay aware and ride defensively, we both knew there was no way I could guarantee a safe return. Thankful to have survived the episode, I remounted tentatively and continued cautiously.

The snow eased as I completed my Bronx tour, steered onto Randall's Island, and crossed the East River into Manhattan at 102nd Street. Bedding down on the Upper East Side after a bit of birding in Central Park—my list pushed to 115 species with the inclusion of Brown Creeper and Common Grackle—I hoped my Manhattan exit would unfold with less drama.

I crossed the George Washington Bridge the following morning and continued south through Fort Lee and Union City on the New Jersey side of the Hudson. Gazing east across the river at Manhattan, the urban hive appeared peaceful. I couldn't hear taxis honking or jackhammers drilling, and I was too far removed to appreciate the art openings, sporting contests, trendy nightclubs, and market surges that make New York one of the most exciting cities on Earth. All I could see from my vantage was a hardened skyline of steel, stone, and glass, an enduring testament to progress, power, and prestige.

I wondered if enough would ever be enough, with everything in New York—and, more generally, in the United States—under unyielding pressure to be bigger, better, and faster, Big Years included. I'd often felt that possession and status had supplanted kindness and compassion as cultural capital across my lifetime, so I was curious if fundamentals like homemade cookies and conversation with our elders still had a place in the American conscience. Rolling counter to an accelerating culture, my two-wheeled transport slowed me just enough to realize what I'd been missing.

Continuing through Hoboken and Jersey City, I steered onto a bridge reaching across the Hackensack River. The span featured a protected footpath on its north side, and I powered up the gentle incline while taking in views of the surroundings.

Absorbed in that distraction, I slammed on my brakes when I noticed a large object blocking my path. Skidding to a stop fifteen feet short of it, I realized it was a queen-sized box spring. It was an unlikely barricade, a poor tie-down job likely allowing it to take flight from the roof of a speeding car. I dismounted and tried to move it aside. Unfortunately, it had landed with such force that it was lodged between the concrete barrier on my left side and the protective railing on my right. Unable to budge the splintered roadblock, I set to breaking it apart with karate-style chops and kicks. It was a prickly process, but I slowly cleared an aisle through which I squeezed my unloaded bicycle. Panniers fetched and reattached, I continued west.

Ten minutes later, a flat rear tire brought me to another halt. It was my inaugural puncture, and I pulled into an industrial park to repair it. I detached the panniers, flipped the bicycle over, and removed the rear wheel, but I had a hard time prying the tire from the hub. That task accomplished after several attempts, I removed the punctured tube and discarded the offending glass shard so it wouldn't cut the replacement tube as well.

The new tube inserted, I tried to stretch the tire back onto the hub, but no matter how hard I pulled from one side and tugged from the other, the rubber resisted. It was like the thing had shrunk two sizes in twenty minutes. Each time I thought I was close to securing the tire, it popped off the rim somewhere else, my frustration festering with each failed attempt. After a half hour, I had only aching arms and frostbitten fingers to show.

I was overjoyed when, on my umpteenth attempt, the tire finally snapped onto the rim. Boss status restored, I reached for my hand pump but was unable to inflate the tire because my extended fumbling had twisted the underlying tube. Realizing I would need to start over, I shouted into the frigid air, "Fuck this shit!"

I hurled my tools across the parking lot and kicked at my panniers; an inadvertent strike on the bike frame sent pain shooting through my foot. I grabbed the incorrectly assembled wheel and ripped the tire from it in a single motion. Raising the rubber loop above my head, I thrashed the ground with it a dozen times, my meltdown ending only when my shoulder felt it might dislocate.

Out of breath, I dropped the deflated tire on the ground and walked away from my bicycle. I had zero desire to continue, but my stroll through the vacant lot helped me regroup. I realized the tire was a solvable problem, so I collected my scattered tools and returned to the bike. Prepared to sink hours into the task, I calmly straightened the tube and stretched the tire onto the rim. The wheel inflated and the bicycle reassembled, I faced the southwestern headwinds, which had strengthened over my dual delays.

Those heavy breezes made my traverse of Newark an exhausting chore, and I struggled to maintain momentum as I moved along the periphery of Newark Liberty International Airport and into Elizabeth and Rahway. Nightfall was approaching, and I feared I wouldn't reach my South Plainfield destination before dark.

I reached my motel just after sunset. I'd covered fifty-three miles, a new daily high, but the crippling cold, shattered box spring, flat tire debacle, and afternoon wind had beaten the hell out of me from start to finish. I took a shower, turned on the television, and settled in for the evening news. The broadcast led with the date: January 19th, 2014. Distracted by morning meditations and afternoon obstacles, I'd forgotten the anniversary of my sobriety. It was exactly four years since my last alcoholic drink—or, more accurately, since my last twenty-something alcoholic drinks. That last of countless blackouts was the terminal punctuation mark on a fourteen-year romance with alcohol and drugs, a mutually abusive relationship that started as a teenage fling.

COOKIES AND CONVERSATION

My alcoholism began on February 3rd, 1996. I was a junior at Hotchkiss, the Connecticut boarding school where I spent my final three years of high school. Five friends and I had been granted permission to spend the weekend in Greenwich, a large house party the featured event. We'd embarked on similar excursions previously, but I'd not been motivated to try alcohol on those occasions. Long-standing curiosity and light-hearted peer pressure finally got the better of me on that February night, and I reached for my first drink, a foamy cup of beer dispensed from one of several kegs.

My uncontrolled drinking over the next few hours was a hugely successful experiment. I proved dominant at beer pong, laughed with my intoxicated buddies, and spent the end of the night sucking face with a random girl with whom I'd been flirting. It was a blast. Alcohol disinhibited me, grew my gregarious personality to new levels, and lent color to what suddenly felt like a very a black-and-white sober existence. Drinking was bliss, and I fell in love with intoxication at my first exposure.

Beer supplanted birds as my primary extracurricular interest in the ensuing months, and getting hammered at off-campus parties quickly became part of my schtick. I possessed zero ability or desire to moderate my consumption, but I was always friendly and never belligerent, circumstances which guaranteed continued inclusion.

Fortunately, several peculiarities of Hotchkiss restricted my burgeoning romance with alcohol. With four to five hours of homework each night, mandatory athletic participation, a half-day of classes on Saturday, and additional community responsibilities, there wasn't much unstructured time left for drinking. The school was located on a sprawling 550-acre campus, so the nearest liquor store was more than a mile away, the proprietor on the lookout for underage buyers. Cars were forbidden for students, so buying farther afield was difficult. Some students smuggled booze onto campus after signing out for the weekend, but expulsion was the standard for any student caught with alcohol. I was unwilling to assume that risk, so drinking was limited to off-campus occasions, usually twice a month through the remainder of my junior and senior years.

Despite those infrequent but intensifying occasions, drinking was a minor footnote on a formative prep school experience. Wide offerings in mathematics, chemistry, biology, and physics set me on my scientific course, and the track and cross-country teams provided athletic opportunities. By my senior year, I captained both of those squads and served as a dormitory proctor, a privilege reserved for those seniors whom the faculty deemed the strongest role models for the younger students. As a final coup, I was elected student body president, and my early admission to Stanford validated everything for which I'd worked during my Hotchkiss tenure. Against those achievements, sporadic binge drinking hardly mattered—or so I thought at the time.

The ensuing summer witnessed increased drinking and an introduction to marijuana—another successful experiment conducted in my best friend's childhood treehouse—but alcohol and drugs didn't grow from a distraction into a focus until I arrived at Stanford. There, growing insecurity would steepen my condition in unexpected ways.

Snug in my motel in South Plainfield, New Jersey, my body aching from the day's ride, those Hotchkiss years seemed another lifetime. If anyone had told me at age eighteen that I'd be a recovering alcoholic and an unemployed, transcontinental cyclist at thirty-five, I would have laughed in their face before pounding another beer. I was young and naive, and my limited worldview couldn't have grasped my future even if it had been revealed to me. I was just trying to have fun before adult responsibility put the kibosh on drinking games, sexual experimentation, and late-night pranks. I didn't understand the grip my nascent alcoholism already had on me, and I couldn't foresee the problems it would eventually create. Four years of sobriety had offered much perspective, and I hoped the bicycle would lend additional insight through the fifth. Curious how the rest of my journey and my life beyond it would unfold, I fell into a deep sleep, the day's misadventures fading against my strengthening sobriety.

SIX

Family Ties

Suddenly and unexpectedly midair, the world seemed a cold, gray blur. I didn't know if I was falling or floating, but I knew I was screwed either way. My back hit the ground with a painful thump, and I saw my bicycle hurtling toward me when I opened my eyes a fraction of a second later. Unable to avoid the collision, I braced myself as the seat and the seventy-five pounds behind it struck my chest. The blow sent a surge of pain through my torso and left me gasping for air. Clutching at my chest and writhing on the frozen earth, I feared Princeton, New Jersey, was the end of my Big Year story.

My breath slowly returned, and I sat up to comprehend my crash. I didn't see another body or bicycle, so it appeared that I was the only person involved. No log, branch, or rock obstructed the path, and I didn't see ice on which I could have skidded. Likewise, the absence of a carcass minimized the possibility I'd run over a squirrel, raccoon, or opossum. A deliberate attack was unlikely, but I couldn't exclude that possibility from my disoriented vantage.

Turning to my bicycle, I saw that the front rain fender was shattered. I also noticed a fresh skid mark six feet away, the short type that indicates extreme deceleration. Putting those seemingly disparate pieces of evidence together, an explanation took shape. If a rock had kicked-up off the earthen path and lodged between my front tire and fender, then the wheel would have jammed. The front end of the bike would have skidded to a near-instant stop, I would have vaulted over the handlebars, and the still-moving rear end would have assumed an airborne trajectory from which to crush me, an absurd possibility I recognized only because I had declined the suggested installation of more expensive, breakaway fenders designed to prevent it.

That economic lesson learned in the most painful way, I dusted myself off and stood up. Two of my panniers had been thrown, so I gathered them before detaching the remaining pair and dragging my crippled bicycle off the path. My bike propped against a tree, I carefully removed the rest of the shattered fender without inflicting additional harm on the wheel. Quick tests revealed the shifters and brakes were functional, so I reattached my belongings and continued.

A bit of riding revealed that my body hadn't survived the crash as well as my bicycle. Beyond general upper-body tenderness, I experienced shooting pain with each inhale, the stab emanating from exactly where the bike seat struck my chest. The pain didn't prevent me from pedaling, and I blogged about the accident after continuing to Trenton for the night. I referenced the chest pain in my post but did so only in passing because I assumed it would be transient.

When ibuprofen and acetaminophen failed to dull the ache overnight and my tossing and turning made sleep impossible, I realized my injury was more serious than I first thought—but not so crippling that I wanted to pause my journey to have it examined. A foot of snow was forecast to fall through the following afternoon, and I wasn't keen on delaying just short of my parents' house in Philadelphia, a destination I hoped to reach ahead of the approaching storm. I would, however, need to conceal my injury once I arrived at their home. If I hinted that I'd likely cracked a rib three weeks into my adventure, it would have caused additional friction between me and my parents.

When I revealed my bicycle Big Year intention to Mom, she wasn't pleased. "This is a joke, right? You're going to throw your career away to ride a bike and look for birds? This is nonsense," she said.

Mom saw life through traditional lenses of career and family. Anything which didn't further those ends was an unnecessary diversion, and her greatest fear—above burglars making off with her valuables, Jehovah's Witnesses knocking on her door, and neighbors' trees falling on her house—was that I'd suffer financial ruin and return to occupy her basement against her will, engaging in periodic raids on the pantry and refrigerator. My aunt had sanctioned her son's twenty-something soul-searching only to end up with a forever-enabled dependent, and Mom couldn't move beyond that doomsday example, despite the glaring differences between her sister's child and myself.

I anticipated her resistance and replied calmly, "No, Mom. I know it sounds crazy. But between staying in my lab, moving into biotech, teaching high school, or selling out to big business, nothing feels right. I need some space, some perspective."

She exploded, "But you're going to get yourself killed! Or paralyzed! I'm not going to take care of a vegetable. You're Sonia's problem from here out—assuming she doesn't leave you over this. I know I would!"

She hung up without saying goodbye. Her words were harsh, but I knew her concern was rooted in love. No mother wants her child, even as an adult, to make regrettable decisions or place themselves in unnecessary danger. I knew Mom cared, but she couldn't have articulated her love less effectively. Her words stung as I digested our exchange.

Mom and I didn't speak for two months. She relayed the conversation to my father, who, in his predictably English manner, dismissed my plan as "complete rubbish" when I spoke with him a few days later. He was willing to engage me in subsequent weeks, but we avoided discussing my undertaking and stuck to the mundane details of our daily lives. While Mom's disapproval was active, Dad's was passive; for him, ignoring my looming project meant it wasn't happening.

Born on different sides of the planet—my mom to a working-class clan in Philadelphia and my dad to a British family in colonial India—my parents shared commitments to education, employment, and family. Mom attended Syracuse University before earning a master's degree in education from Temple University and teaching in the Philadelphia public school system. Dad immigrated to the United States after graduating from Oxford University and became a very successful lawyer without attending law school, a feat he managed after passing the

Pennsylvania Bar Examination on his own volition. They met at a party, eventually married, and brought me and Imogen, my younger sister, into the world a few years later.

More pragmatists than dreamers, Mom and Dad saw my Big Year as a loss of focus, the journey's personal possibilities insignificant against its professional and physical pitfalls. Though they'd voiced strong opposition in the months leading up to my Big Year, I hadn't heard much from them since getting underway. That lack of communication was a relief because I didn't have the time or energy to engage their disapproval on top of biking, birding, and blogging. They did invite me to stay with them when I passed through Philadelphia, but I wasn't sure what to expect as I raced the winter weather toward their house. Two inches of fresh snow coated the driveway when I arrived on the afternoon of January 21st, barely twenty-four hours removed from my rib-cracking crash.

The front door was locked when I arrived—recall Mom's chronic fear of burglars—so I rang the bell. I heard a series of locks open before she emerged with an unanticipated smile on her face.

"I'm so glad you made it! This weather is awful," she said. "I'll open the garage so you can stash your bike."

I did as instructed before trying the back door, but my face smashed into the glass pane when I tried to open and walk through that also-locked portal. Mom heard the thump, unlocked the door a moment later, and hugged me as soon as I stepped inside. Her squeeze hurt like hell, but I bore the pain silently for fear of revealing my secret. Dad emerged from his study and offered another excruciating embrace.

"Blimey! You smell dreadful!" he said. "We've made you a late lunch, but you can't eat until you shower."

I'd expected our interaction to be awkward and forced, so I was surprised by their upbeat attitudes. I cleaned up and joined them in the kitchen twenty minutes later.

"How do you feel after yesterday's accident?" Mom asked as I reached for some pasta salad.

I was caught off guard because I had assumed they wouldn't know about it. I conjured a quick lie to keep them from worrying. "A bit sore.

I don't think it's anything major," I said, my left side smarting as I spoke. "How'd you hear about it?" I asked.

"On your blog," Dad said.

What the hell? Had I entered a parallel universe where my parents actually cared about my Big Year?

I spoke disbelievingly, "You read my blog?"

"Every day. We missed a few at the start but caught up later. It's great how people are chiming in from all over the country," Dad replied.

Mom added, "And I share all your entries on my Facebook page."

Shut. The. Front. Door. Had my Luddite parents graduated from the Mark Zuckerberg School of Social Networking since we last spoke?

"You know how to share stuff on social media?" I asked.

Mom replied, "Oh, yeah. You're a big hit with all our friends. Everyone wishes they'd done something similar before getting bogged down with careers and kids."

"We're just amazed at the interest and support you've generated," Dad added. "It seems like the whole birding community is behind you. It's incredible."

I'd made a final plea for them to understand the journey's potential during my Christmas visit a month prior, but they had resisted on every front. Apparently, the project needed to be in motion before they could appreciate it, the enthusiasm of others inspiring my parents to find their own. It was a remarkable attitude adjustment, and I was ecstatic that they were finally behind me. I had hoped my self-powered journey would change my perspective, but I never expected it would change theirs as well. That they'd finally come around made me feel like anything was possible.

Mom interrupted my thoughts. "So, where do you go from here?"

We spent the remainder of the afternoon discussing my plans. Six inches of snow had fallen by dinner, and the forecast indicated that heavy precipitation would continue through the night. The roads were buried beneath a foot of accumulation the next morning, so I conceded the day as an opportunity to spend time with my newly positive parents. Supportive as they were, I didn't elaborate on my ribs; however badly the injury hurt me, I wouldn't risk allowing it to upend the tenuous understanding we'd forged.

The roads were cleared by midday on January 23rd, so I resumed my southbound course. My parents were sad to see me leave but excited to follow the rest of my adventure. Dad gave me another painful hug and reminded me to be careful. Mom shed many tears, but it was reassuring to know they represented love and support instead of anger and frustration. Through her sobs she whispered a rare "I love you" in my ear. With that sentiment usually more implicit than explicit in our family, her words left me emotional as I rolled down the driveway and out of her view for another eleven months.

After an overnight in West Philadelphia, I cycled to the urban oasis of John Heinz National Wildlife Refuge to look for Northern Shrike, a gray, robin-sized bird displaying a mischievous black mask across its eyes. Shrikes are evolutionary oddities because they are a rare example of a predatory songbird. Unrelated to birds of prey, shrikes hunt rodents and small birds, as do hawks, eagles, falcons, and owls. Shrikes cannot kill their quarry with raw strength as those raptors can, so they've evolved the ability to impale their catches on snapped branches, large thorns, or—with the rise of agriculture and ranching—barbed-wire fences.

One of only two New World examples, the Northern Shrike nests in boreal forest across Alaska and Canada and—similar to other high-latitude specialists I'd sought in my first three weeks—reaches into the northeastern United States in only small numbers each winter. I planned to search for the bird in Massachusetts, but the blizzard barred me from several of its usual haunts. Rather than waiting for access to those areas, I continued south on the assumption I'd bump into a Northern Shrike somewhere else in New England. When that didn't happen, I pinned my hopes on one Philadelphia individual, the faithful bird reported daily since its mid-December discovery. With no recent sightings to the south, the odds suggested that the Philly shrike would be my last chance at the species.

When I arrived mid-morning on the January 24th, the refuge was buried beneath a foot of snow and the massive impoundments were iced over. Eager to keep my blood flowing, I ditched my bicycle at the visitor's center and began the mile-and-a-half slog to the back of the refuge, the area the shrike had favored during its month-long sojourn. With the snow reaching my knees, I worried that the bird had departed for lack of open hunting grounds, but my fear evaporated when I spotted the masked assassin at the top of a leafless tree forty minutes later. Oblivious to the Arctic conditions, the enduring bird surveyed the surrounding meadow without paying me any attention. Its ability to outlast the cold exceeding mine, I scampered back to my bicycle and continued to my place of lodging for the night.

The shrike was my final cold-weather target and capped an incredible start to my bicycle Big Year. I tallied only 120 species through my first 24 days—I expected to see that many in a day once spring migration began—but the shrike and the other high-latitude species were particularly valuable from a Big Year standpoint because they would be absent from my remaining route. With those twenty-something northern species secured, I had a mathematical path to 600—assuming I survived another 11 months and 14,000 miles of riding.

My birding triumphs were significant in isolation but unbelievable against the continuing polar vortex. Daily temperatures hovered in the teens and twenties, and the mercury moved above freezing on just one day of my first twenty-four. Five independent storms had complicated my progress, and I'd lost four riding days to snow and ice: two in Massachusetts, one in Rhode Island, and one with my parents in Philadelphia. Despite those delays, I'd cycled 665 miles over 20 usable days, averaging 33 miles per day. That wasn't much daily distance, but it seemed as much as could reasonably be expected given my cycling inexperience, the atrocious conditions, and overarching birding demands. My bicycle Big Year was the marathon of marathons, and I needed to pace myself early to avoid burnout later.

Minus my chest injury, I could not have been happier to have reached Philadelphia, a destination I'd silently imagined as my fail-safe point. If I realized my undertaking was a mistake by the time I reached that familial port, then I could quit the bicycle and formulate an alternative life plan from my parents' house—assuming they let me in.

That worst-case scenario avoided, my northeastern leg was a successful prototype on which I could model the rest of my journey.

Temperatures would climb as I cycled south, longer days would allow more birding, and the promise of new species would provide motivation as I powered toward Florida. Spurred south by more falling snow on January 25th, I cranked out of Philadelphia and toward Wilmington, Delaware, my gateway to warmer weather—or so I hoped.

SEVEN

Southern Migration

In 1763, surveyors Charles Mason and Jeremiah Dixon were charged with resolving a boundary dispute between three independently chartered English colonies: Pennsylvania, Maryland, and Delaware. The task required political navigation, but the pair fixed the borders of the eventual states in 1767. Among their delineations, the Pennsylvania–Maryland boundary assumed particular significance because it later marked the northern limit of institutional slavery. The Union victory over the Confederacy in the Civil War rendered that institution moot, but the Mason–Dixon Line has endured as a symbolic demarcation between the North and South.

Growing up twenty miles north of the line, in Philadelphia, I noticed it also partitioned the harsher winters of points north from the milder spells of places south. My New England start presented challenges I hadn't imagined, but the envisioned Mason–Dixon relief motivated me as I battled bone-chilling conditions through New York, New Jersey, and Pennsylvania. As long as I didn't freeze to death short of that line, my childhood experience suggested conditions would improve south of it.

My presumption might have held in an average winter, but the continuing polar vortex in early 2014 offered no relief. The Mason–Dixon was covered with ten inches of snow when I crossed it on January 26th, and temperatures in the low-twenties suggested Maryland was closer to

Canada than Dixie. I was heavily bundled at a latitude where I expected to be shedding layers, and I couldn't utilize biking shoes, the sort that clipped to my pedals and generated upstroke force, because it was too cold for uninsulated footwear. Until conditions improved, I'd struggle along in my winter boots.

Considering biking and birding, the best route through Maryland and Virginia would have been along the scenic eastern shore of the Chesapeake Bay and across the twenty-two-mile bridge–tunnel at the mouth of that body of water. Bicycles are prohibited on that span, so I was forced south along the developed western shore instead. I followed US Route 1 through the downtown depths of Baltimore and Washington, DC, the latter navigated under falling snow, and reached Alexandria, Virginia, on January 28th. Temperatures on that day and the next topped out in the teens, and I wondered if the polar vortex would ever loosen its icy grip on me.

Conditions improved as I powered south through Virginia. Overnights in Dale City and Fredericksburg delivered me to Richmond on January 31st, and I debuted the biking shoes when the mercury soared to forty degrees south of Petersburg on February 2nd. Wheeling through seasonally fallow cotton fields, I pulled over and fished the pair out of my panniers. Black sneakers with metal plates in the soles, they would clip to my pedals and allow me to pull with my hamstrings in addition to pushing with my quadriceps. I emancipated my feet from my boots, plunged them into the shoes, and strode along the shoulder, the metal plates clicking with each step. I needed several wobbly starts to practice clipping the plates into the pedals, but the energetic benefits manifested immediately, the Virginia air whooshing across my ears as I raced through the agricultural landscape.

Cranking south toward the North Carolina state line, I spotted a Wilson's Snipe, a long-billed sandpiper that prefers fields to beaches, poking around in a roadside puddle. I braked for a better view of the brown bird but found my shoes clipped to my pedals when I tried to casually dismount as I'd done for the previous month. Without the gyroscopic force of spinning wheels to counterbalance my sudden flailing, I toppled left into the road.

With my bicycle pinning my left leg to the tarmac, pain coursed through my body. My contortions suggested a giant rodent caught in an oversized mousetrap, but I yanked my leg free and slithered out from under the bike.

My ribs returned to their baseline ache, but a scraped left wrist required attention. Cleaning the wound, I couldn't help but laugh; viewed from another perspective, my flop would have been entertaining slapstick. Remounting, I hoped I could remember to unclip before stopping in the future.

I didn't, and inexperience struck a second blow when I braked to investigate a sparrow, again without unclipping. Realizing my mistake, I conceded a rightward spill to protect my aching left side and toppled into a ditch filled with a suspicious sludge that looked like a chocolate slushy and smelled like dog shit.

What the hell am I doing? I left my career for this?

Zero humor in my second tumble, I righted myself and sloughed off handfuls of half-frozen nastiness. I dragged my bike out of the ditch, gruffly clipped in, and continued.

My third fall came under a soaring raptor and mirrored the first with a leftward flop. Seasoned at asphalt diving, I cushioned the impact with my forearm while shaking my feet free. My ribs throbbed, but rage quickly outstripped pain. I popped up, flailed my arms, and exploded, "Fucking shoes! You ain't shit! And bike? Fuck you, too, you son-of-a-bitch! I'm over this shit!"

I ripped off my helmet and launched it into the cotton stubble before addressing the shoes. I tore off the Velcro straps but was stonewalled by double-knots. Instead of fighting through those, I yanked the tied shoes off and propelled them into the same field.

Exhausted by my fit, I plopped onto the grassy margin. Seated on my ass, legs extended in front of me with two mud-stained socks staring back, I felt defeated, mine an epic fail. Shoulders sagging, I sulked, a pair of crows cackling at me from a nearby phone pole.

My falls were maddening, but I eventually cooled off and recognized their inevitability. Even if I'd mastered the bicycle shoes before departing, a month in the boots would have erased my procedural memory.

Given that realization, I was thankful my initiation had unfolded in the middle of nowhere; had I been riding a busier road, scrapes and bruises would have been the least of my worries. Solace gleaned, I retrieved my gear before replacing my shoes and helmet and clipping in for the fourth time. That iteration proved the magic number, and I crossed into North Carolina without additional incident.

My motel, the only one in Murfreesboro, was as disastrous as the day's ride. I'd seen crack houses with more curb appeal. The window screens were rusted and ripped; weeds protruded from every crack in the decaying pavement. The lock to my assigned room was formality; the rotten plywood door would have yielded to a modest throw of a hip.

Any hope my lodging would play better inside vanished when I opened the door. A lightbulb dangled from exposed wires, and the walls were stained brown from decades of smoking. With sticky shag carpeting clinging to my shoes, the place was somewhere between a homemade porno film and a heroin overdose. The bathroom was no better, mold suggesting the shower to be a biology experiment. Unable to face that petri dish, I used the sink and a washcloth to clean myself of mud and blood.

Returning with a take-out dinner from McDonald's, I spied a group of people lurking behind the motel. I'd been involved in enough shady situations to know when drugs deals were going down, so I walked the long way around the building to reach my room. The last thing I needed was a group of cracked-out meth fiends taking interest in me.

I watched the Seattle Seahawks trounce the Denver Broncos in Super Bowl XLVIII and sank into the ramshackle bed when the contest concluded. A heated argument in the parking lot unsettled me—someone owed someone money, presumably for something other than Girl Scout cookies—so I pushed the chest of drawers in front of the door and placed my utility knife on the pillow next to me. With that insurance, I drifted off, residual hoots and hollers minimized against the day's falls and flops.

Beyond the Hell Motel, Murfreesboro wasn't notable among the small towns I'd visited since crossing the Mason–Dixon. Dated main streets lined with abandoned storefronts suggested most communities were economically depressed, and—in one extreme example—a dilapidated Goodwill was the only operational enterprise. Big-box retailers and fast-food chains had extinguished locally owned businesses, Walmart being the beacon in many municipalities, and I could feel the economic lifeblood flowing out of the faltering communities and into the pockets of faraway shareholders. It was depressing: globalism, greed, and supply-side vultures feasting on what remained of small-town America.

Departing Murfreesboro, I cranked out eighty miles to Greenville on February 3rd and another fifty-one to New Bern on the 4th. Following Highway 17 across the Coastal Plain of North and South Carolina, I intersected the Atlantic north of Wilmington and reached Myrtle Beach, South Carolina, on February 9th. Highway 17 felt endless, but the appearance of southeastern species like White Ibis, Brown-headed Nuthatch, and Red-cockaded Woodpecker suggested progress.

The weather had cooperated since I left Murfreesboro, with daily temperatures in the forties, but a reenergized polar vortex thrust frozen rain onto South Carolina as I rolled into Mount Pleasant on February 11th. The Ravenel Bridge, my connection to Charlestown and points south, was closed to all traffic, bicycles and pedestrians included, after massive chunks of ice had dislodged from the towers and crushed several cars. That closure pinned me for two days, and I lost two more to thirty-mile-per-hour southern headwinds as the vortex contracted north. Exploring beaches and marshes on Sullivan's Island during the delay, I added the elegant Black Skimmer and the adorable Piping Plover to reach 179 species.

Conditions improved on February 16th, and I reached Savannah, Georgia, on the afternoon of the 18th. The most charming of all the southern municipalities I'd visited, the city's sleepy streets were framed by low-slung oaks and the Spanish moss which dangled from them. Porch swings invited daydreaming, and I imagined Savannah a wonderful place to grow up. Among new birds that appeared on the Georgia coast, four species of sparrows—Nelson's, Saltmarsh, Henslow's, and LeConte's—brought my total to 193 species.

Though the Southern hospitality was wonderful, my hosts stuffing me with biscuits, fried chicken, and macaroni and cheese, the trademark courtesy did not extend to the road. I'd been warned that Southern drivers are unwilling to accommodate cyclists, but I wasn't prepared for the hostility I experienced in Virginia, the Carolinas, and Georgia. Combative horn honking was routine, and I was often given little leeway when more space was clearly available. Good ole boys hurled insults from passing vehicles—"Git off the road, hippie mother fucker" and "Fuck you, faggot cocksucker" representing the pinnacle of their creativity and obscenity—and I was struck by bags of fast-food trash on two different occasions.

My most memorable confrontation unfolded in South Carolina. I was cruising along the shoulder of Highway 17 when I heard a very loud vehicle approaching. Glancing over my shoulder, I saw a tricked-out pickup truck overtaking me in the left of two traffic lanes. The passenger offered a nefarious grin, and a disconcerting possibility materialized when I saw their huge chrome exhaust: they were going to try to "roll coal" on me.

To roll coal, a driver modifies their engine to emit clouds of black smoke from the tailpipe, usually on demand. The deranged display was conceived by climate change deniers to give environmentalism a public middle finger—to what end, I'm not sure—and cyclists and hybrid drivers are favored targets because of their perceived concern for the planet.

I'd been on the lookout for such braindead bullies, and my suspicion was confirmed when the driver aggressively maneuvered the growling monster truck into the right lane ahead of me. Their plan was transparent, so I dipped onto a side road, avoided their pollutant-laden plume, and escaped under another hail of insults. Against the effort I was making on the bicycle, their environmental disregard was revolting.

I survived two treacherous weeks on Highway 17 and crossed into Florida on February 22nd. I experienced a tremendous sense of accomplishment, nearly 2,000 miles covered since I left Sonia in Massachusetts, but I had another week of riding to reach the best birding at the southern end of the peninsula. Following State Road A1A to St. Augustine, I added Yellow-throated Warbler for species #200, a milestone deserving of a slab of blackened snapper in Daytona Beach. My daily workouts on the bicycle licensed me to eat whatever I wanted, ten pounds already shed, so I backed the fish up with a slice of pizza and a triple scoop of ice cream.

Strategizing in Daytona, I learned of a Clay-colored Sparrow in Titusville. The species occupies the Midwest and Great Plains—a sighting was all but guaranteed in Texas—but the Titusville example was an opportunity to eliminate it from my thinking. The bird had frequented the same backyard bird feeder for a month, so I used my contacts to arrange a visit with the homeowner.

I met Bernie at his home midday on February 26th. Gray-haired and tanned, he was in his late fifties or early sixties, a scraggly mustache framing an imperfect but welcoming smile. He invited me into his modest abode, and we settled at the kitchen table, where we had a view of his bird feeder.

"This'll be more comfortable than your bike," Bernie joked as he pushed a padded chair in my direction.

"Thanks, man. A sore ass is pretty standard by now," I replied.

"The bird cycles through every twenty minutes or so, but you're welcome to stay as long as it takes," he said as he handed me a glass of ice water.

"How many people have come through for this bird?" I asked.

Tilting his gaze skyward while tabulating, he responded, "Fifteen, maybe twenty. All from Florida. The bird has shown for everyone, so don't you be the jinx!"

I chuckled, our conversation continuing as we waited for the sparrow. My travels provided conversational fodder, but our exchange swung toward the recent passing of Bernie's wife.

"I loved her so much. She was my angel, my everything, and now she's gone," he said with a soft sob. "I'm just not sure I can go on without her."

I replied, "I'm sorry you didn't get more time with her. That's all we want, right?"

"I just feel robbed," he said, looking at a wall of framed pictures. "These are all I have now."

"I'm sure you have tons of other memories, too. There isn't a wall big enough for those. They're your complete picture of her, and nothing can take it away."

Bernie straightened his back and spoke. "Yeah, that makes sense. And I'm sorry to put this on you. It's just good to have someone to talk to."

The Clay-colored Sparrow flitted onto the bird feeder in the next moment, and Bernie rebounded while offering a chronicle of the bird's month-long stay. Cross-referencing that history with his eulogy, I surmised that the sparrow had appeared two or three days after his wife passed and had, somewhat symbolically, visited daily since. I wondered if birders had helped Bernie cope with his loss. Some people keep news of rare birds on their properties quiet for fear of having their space invaded by overzealous birders, but Bernie had welcomed all comers. I didn't know his motivations, but I was honored he'd invited me into his home and allowed me to serve as a confidant during a difficult time.

Bernie and I talked birds for half an hour beyond the sparrow's reveal, but I still had forty miles to go to reach my Melbourne destination. We wound our conversation down, and I thanked him for his hospitality. He walked me out to my bike at the side of the house.

"Life's like a year on a bike: slow and steady," I said. "I can't ride the hundredth mile until I've finished the first, right? Let's both keep going and see where we end up."

He cracked a smile and thanked me for visiting. With a pat on the back, he sent me down the driveway, onto the street, and back to Route 1.

I spent the afternoon thinking about Bernie and the sparrow. It was then late February, and the immutable desire for birds to migrate north, to their breeding grounds, would soon prevail. Bernie's sparrow would be no different, and I wondered what the bird's departure would mean for him. Visits from birders would cease, and he would be left to grapple with his grief alone. I was sure he would manage, but the thought of him suffering in silence was difficult. Never one for prayer, I uncharacteristically asked that the sparrow stay in Bernie's yard forever. As long as it did, his wife would live on—at least in conversation.

EIGHT

My Shadow

Gazing over the wetlands, I spied a charcoal raptor riding a rollercoaster of sweeping dips and airy bounds. He was hunting, his gaze fixed downward as he surveyed the vegetation. When something caught his eye, he wheeled, hovered, and darted into the reeds. Temporarily disappeared, he reemerged with his orange foot wrapped around a snail. He flew to a snag and used his hooked beak to extract the tender morsel from its shell, his finesse suggesting a sommelier uncorking a vintage Bordeaux. The mollusk consumed in a single swallow, the predator kited into the marsh to find another slow-moving snack.

Snail Kites inhabit freshwater wetlands from Mexico to Argentina with a distinct subpopulation in Florida, those birds possessing smaller beaks than their Latin American counterparts. Human encroachment and associated development caused the Florida birds to decline through the early twentieth century, and the species—then called Everglade Kite—was listed as endangered in 1967. The birds have partially recovered under that protection, but they and their freshwater habitats face continued ecological pressure on several fronts.

The biggest threat is human population growth. Fewer than two million people lived in Florida in 1940, but that number had ballooned to twenty million by 2014. The more acres humans occupy, the fewer that remain for Snail Kites and other species.

Water management is another concern. Increased levels of engineering have been imposed to ensure that developed areas neither drown nor dry out, and that network of dikes and canals has disrupted the peninsula's water cycle. Florida's Snail Kites have historically sustained themselves on the Florida apple snail, an aquatic invertebrate that is sensitive to water levels, so human manipulation of the water table has adversely affected the birds through their finicky food source.

Non-native species constitute a third problem. Humans transport species from one part of the globe to another, and Florida's climate allows many escaped non-natives to thrive, often at the expense of resident species. Burmese pythons are decimating bird populations; feral hogs are trampling sensitive habitat; ornamental vines are choking out local vegetation. Even a non-native snail can cause ecological reverberations.

Widespread in Latin America, the lemon-sized island apple snail is a relative of the strawberry-sized Florida apple snail on which Florida's Snail Kites feed. The island apple snail was introduced to Florida when people dumped aquariums into public waterways, and the non-native thrived because it's less sensitive to water levels and reproduces faster. Those advantages suggested the island apple snail would outcompete the Florida, and biologists feared that outcome because Florida's Snail Kites have evolved smaller beaks to eat the smaller snails.

Against those predictions, the birds have demonstrated a surprising ability to prey on the larger snail. Kite numbers have increased since the island apple snail's introduction, and research has revealed that Florida's kites are evolving larger beaks in response to this larger snail; Florida's kites are looking more like their larger-beaked, Latin American counterparts, which evolved to eat the island apple snail in its native range.

Despite that encouraging outcome for the kites, the island apple snail consumes vegetation faster than the Florida; it's therefore possible frogs or ducks will suffer from lack of protective cover in the future. Ecosystems don't lend themselves to controlled manipulation, and deleterious outcomes often aren't recognized until the balance has been disturbed and the possibility for remediation has expired.

No species exterminates others as efficiently as *Homo sapiens*, and contemporary extinction rates are orders of magnitude above historical baselines. Ecosystems thrived for millions of years without humans, but we won't last long without healthy habitats. Despite what industry advocates claim, environmentalism isn't misplaced bets on birds, snails, and other organisms; it's a calculated hedge against self-destruction. Watching the Snail Kite disappear into the distance, I hoped management would be enough, for the bird's sake and ours.

I spent the next week moving south through the sprawl of Palm Beach, Delray Beach, Fort Lauderdale, and Miami. I encountered the bizarre Roseate Spoonbill, his whimsical pink plumage softening the hardened appearance of his spatulate beak, and I enjoyed mesmerizing views of the iridescent Purple Gallinule, her elongated yellow toes distributing her weight across each lily pad as she strode from one to the next. Though drabber, the grayish La Sagra's Flycatcher was most exciting from a Big Year perspective; typically ranging through the Bahamas, Cuba, and the Cayman Islands, a vagrant example in Coral Gables was a bonus for species #230.

Climbing into bed in Cutler on March 7th, I readied for my peninsular traverse; with no viable lodging between the Atlantic and Gulf Coasts, I'd need to cover the hundred-plus miles in a single ride. The sun cast a long shadow ahead of me as I powered onto Highway 41 (the Tamiami Trail) the following morning and cranked through the awakening Everglades. Initially distorted, the shadow resolved into bicycle and rider as the sun climbed higher. A swaying carpet of reeds reached to the horizon in every direction, and an army of alligators eyed me as I wheeled through their roadside ranks. Stretching to the horizon, the Tamiami tarmac ushered me west.

Bald cypresses rose from the swamp; buttressed by supporting roots and bearded by Spanish moss, the trees shaded me from the midday sun. I played hide-and-seek with my shadow as it darted between those of the overhead trunks and branches. For one illuminated instant, I existed; in the shaded next, I disappeared, my shadow obscured by those of the arbors. It was a kinetic and playful dance, a two-dimensional representation of my movement through three-dimensional space.

The sun sank westward, and my shadow correspondingly lengthened behind me. I labored out of the trees, pedaled through coastal scrub, and reached the Gulf at Marco Island. By the time I explored Tigertail Beach and arrived at my overnight, I'd churned out 111 miles. I was exhausted, but it was a healthy fatigue, the sort that builds strength rather than causes burnout. My Atlantic winter completed and my Gulf spring initiated, I reflected on my growing cycling confidence.

While planning my Big Year, I estimated I'd need to ride 15,000 miles—about 40 per day—to find 600 species. To Marco Island on March 8th, my calculations held perfectly; I'd cycled 2,746 miles across 67 days for an average of 41 miles per day. Weather had delayed me at multiple points, but those pauses were disguised blessings; the addict in me lacked the discipline to schedule downtime, so Mother Nature assumed that responsibility on my behalf.

Energized by my peninsular traverse, I powered north through Naples, Sanibel Island, Fort Myers, and Tampa in subsequent days. Arriving in Clearwater on March 13th, I was shocked by the numbers of people I encountered. Beaches were packed, bars were hopping, and carloads of twenty-somethings cruised the main drag with their radios blaring. The clues aligned; I'd cycled straight into spring break. Pedaling through the drunken mayhem, I felt like I was in college all over again.

I matriculated at Stanford in September of 1997. My plan was to study a combination of biology and chemistry, and I joined a life sciences track loaded with hypercompetitive premed students. Hotchkiss had prepared me academically, and my familiarity with dormitory living guaranteed an easy transition to the picturesque Palo Alto campus.

I gravitated toward drinkers in my dorm and on the Ultimate Frisbee team because they shared my outgoing and boisterous tendencies, and I formed a network of buddies with whom to drink beer, smoke weed, and clown around. While Hotchkiss punished drinking with expulsion, Stanford's alcohol policy was laissez-faire; with a lack of governance, an initial academic ease, and a posse who liked to party, my drinking exploded. A month into my first semester, I spent more nights shit-faced than sober.

Alcohol made me a presence at every party I attended, and my antics—chugging a pitcher of beer or jumping off a roof—became notorious around campus. I had no desire or ability to moderate my alcohol intake, and anything short of annihilation felt like a half-assed effort at fun. I only wanted to get hammered, do stupid stuff, and laugh about it the following day. Whether it was a speedy skateboard slalom through a six-story parking garage or a midnight plunge from the ten-meter platform in the diving facility, every drunken night promised something unexpected.

Unfortunately, steepening blackouts ensured chunks of time went unaccounted as my first year progressed. I routinely said and did things I didn't remember when I was inebriated, and my assigned roommate, a nice but straight-edge guy, requested a switch after I urinated on the floor in a drunken stupor. The situation resolved when I moved in with a friend down the hall, and I stayed off the institutional radar despite continued binge drinking.

Midway through the year, I was eating lunch with Brynn, who lived on the second floor of my dorm. I first noticed her at freshman orientation, and our subsequent interactions revealed her to be smart and funny. Residential proximity facilitated daily encounters, and we had obvious chemistry despite our dichotomous interests; she was a history major who sang a cappella and rarely drank alcohol. Sweet on her nonetheless, I hoped circumstances would allow us to grow our friendship into something more.

Seated with her in the cafeteria, I was recapping what I could remember about the previous night—something about a beer bong and bicycle—when she interrupted me.

"You're still at it, huh?" she asked with a mix of sarcasm and curiosity.

"You know it," I said. "Even made it to nine a.m. class. Completely unstoppable. You going to go to Sigma Chi tonight?"

"Sure. Totally looking to hook up with drunk frat guys," she said with an eye roll.

I laughed and replied, "Yeah, not your scene. I know. I'm headed to Theta Delta beforehand, so I'll prolly be ruined by the time I roll over there."

"Seriously though, how long can you keep this up? Everyone in your major stresses 24/7 and you're partying all the time. How do you do it?"

"It's easy. I don't give a shit."

At least that's what I wanted everyone to think. An academic, athletic, and community standout at Hotchkiss, I needed just two weeks at Stanford, among National Merit Scholars and future Olympians, to recognize that I wouldn't distinguish myself as I had in the past. Drinking, however, was a different story, and my uncanny ability to stay afloat despite constant intoxication differentiated me from everyone in my major and all but a few on campus. I constructed my image around partying, and I worked incredibly hard at short but consistent intervals to maintain it.

Brynn pressed me. "But don't you think that's sad?" she asked. "I mean, don't you want to make the most of your time here?"

"Sure. But all my classes are graded on a curve. I'm never going to be an A or A-plus student, so I'll take my A-minus and have fun along the way."

That was another half-truth. Beyond motivating late-night antics and facilitating promiscuity, drinking was semi-calculated self-sabotage. Fearing I wouldn't be the best under sober circumstances, I used alcohol to excuse imperfect performance. My grades were still very good, but I experienced a pathetic sort of satisfaction when my classmates wondered, often aloud, how well I would do if I applied myself. As long as I was drinking, no one knew—the imagination thankfully greater than the reality.

Brynn followed up, "And getting hammered all the time is fun because . . . ?"

"It's a way to blow off steam. After this, it's going to be the real world for the rest of our lives. I'm going to party while I can," I replied.

She raised an eyebrow and countered, "And you're not worried about your ability to slow down in the future?"

"Nope," I said. "I can stop whenever. Responsibility will put the clamps on me eventually, so I'm not worried."

She replied, "I'll be curious to see how it goes. For what it's worth, I think sober Dorian would be a pretty cool guy. He might even be worth dating."

My second year at Stanford unfolded identically to my first, but two changes during the ensuing summer, while I was interning with a biomedical research group at Rockefeller University in New York City, had lasting consequences. The first was my emerging interest in developmental biology, the process through which different cells—neurons, muscles, skin, and so on—are specified and organized during embryogenesis. Stanford coursework seeded my curiosity, but my first taste of experimentation at Rockefeller left me wanting to understand how an organism is built from an intrinsic genetic blueprint.

Infatuated with that problem when I returned to Stanford for my junior year, I initiated an independent research project to explore the genetic specification of neuronal subtypes in the rodent brain. I loved designing, conducting, and interpreting experiments, and I enjoyed the rhythm of lab work even when I was hungover. By the end of that third year, I was committed to pursuing a biomedical PhD, my goal to become a professor and direct a research group at a university like Stanford.

The second change that occurred during my Rockefeller summer was more sudden, manifesting through a nondescript pill a friend handed me outside of the Hammerstein Ballroom, a massive concert venue in Midtown. I knew 3,4-methylenedioxymethamphetamine acted on the same serotonin and dopamine neurotransmitters as the monoamine oxidase inhibitor (MAOI) class of prescription antidepressants, but I wanted to experience the comparative avalanche of those same "feel good" molecules the narcotic elicited.

The sensation started a half hour after I took the pill, just as I was wondering if the drug had been overhyped. Optimism gripped me, and I felt a sudden affection for everyone around me. My skin tingled, my breathing deepened, and my heart accelerated as techno and breakbeats moved me to join a swirling milieu of dancing bodies. Every casual touch was electric, and the collective emotion grew with each track the DJ dispensed. Beneath sweeping lights, flashing strobes, and fog, time compressed, my introduction to ecstasy ballooning into an all-night euphoria. The pharmacological effects abated after sunrise, but the experience resonated for much longer.

Enthralled with dance music and club culture after attending that first rave, I returned to Stanford, bought a pair of used turntables, and—when I wasn't studying, playing Ultimate Frisbee, or teasing apart neuronal subtypes in the lab—taught myself how to mix records. Honing my craft through the fall, I found myself booked most weekends through the second semester of that third year. At $100 per hour and free booze to boot, I was living the alcoholic's dream.

Although drinking relegated birding to an occasional distraction during my first two years at Stanford, my coincident discoveries of developmental biology and DJing put pay to the interest. On the academic front, when I was in college there were few similarities between ornithology and developmental biology. Generally, organismal fields like ornithology, ecology, and evolution relied on observation, while molecular fields like developmental biology, cell biology, and genetics relied on experimentation. I enjoyed generating hypotheses and designing assays to probe them, and my reductionist mind seized on developmental questions over ornithological study.

Those distinctions are nuanced, but the role my DJing played in the demise of my birding interest is more straightforward: it was tough to go birding when I spent my free time mixing records, drinking, and recovering from hangovers. While I once dreamed of spending months in South American jungles, my focus constricted on cities and associated nightlife as my DJ interest took hold.

I graduated from Stanford in June 2001 with an honors degree in biological sciences, a position as a research assistant at Harvard, and an abusive relationship with alcohol, one which had prescribed forty to fifty drinks a week through my senior year. I perceived my behavior as harmless throughout, and I assumed future drinking would taper against post-graduate responsibility. I didn't realize my affliction would contrarily compound over time, my eighty-proof problem to shadow me as I stumbled between parties in the ensuing years.

My reminiscence was interrupted when an intoxicated twenty-something staggered off the beach and into my path. While swerving around him, I shouted, "Heads up!"

He came to a swaying halt and spoke with a slur. "Chill out! Just grabbin' a few more beers, bro."

He stumbled across the street behind me, and I wondered how many collisions I'd similarly sidestepped while at Stanford. Given all the reckless stuff I did across those four years and the decade beyond them, it was remarkable I'd survived without major injury—bumps, bruises, and cuts notwithstanding.

I reached my Clearwater motel in late afternoon, and my anticipation grew as I showered and dressed. Sonia and I had spoken each day we'd been apart, but we hadn't seen each other since I left Massachusetts ten weeks earlier. We'd managed the distance well and coordinated travel plans to meet up in Clearwater, Sonia flying in to Tampa from her temporary Nashville base.

We spent time at the beach, played several rounds of miniature golf, and took in a Philadelphia Phillies spring training game after renting Sonia a bike for the day. In this way, we undertook the commute to the stadium—in line with my self-powered promise—without petroleum. That's the kind of woman she is, willing to accommodate my imagination however ridiculous it was. I'd missed her sarcastic sensibilities and tender touch, and her presence was a poignant reminder of what I'd risked in undertaking a bicycle Big Year. That monumental task beckoning after three days together, we shared a final embrace. Our next meeting remained undetermined at that Clearwater juncture.

I started north after stroking her cheek a final time, a permanent reunion all that I wanted at that moment. She'd served as my lighthouse for six years, and I couldn't imagine my future without her. If I survived another nine months and 12,000 miles, then we'd be together again.

Powering north toward Ocala four days later, that time and distance seemed an eternity as an SUV invaded my designated bicycle lane. Visions of Sonia gripping me, I braced for impact.

NINE

Risk Management

I was cruising through the intersection at fifteen miles per hour when the SUV made an unannounced right-hand turn across my path. I was unable to slow in that split second, and my front tire impacted the side of the vehicle, behind its rear wheel, the collision spinning my handlebars to the right and ramming my left shoulder into the right rear window as the SUV continued through the turn. That support lost in the next instant, I careened into the adjacent lane, collapsed onto the pavement, and braced for a follow-up strike.

I heard tires screech and looked up to see a motorcycle skidding to a stop twenty feet short of my face. A bearded, leather-clad man dismounted and rushed toward me.

He knelt and touched my shoulder. "Jesus Christ! You OK, man?" he asked.

In shock, I stuttered, "I- I- I think so."

"Good. Stay down," he said. "I dunno what that idiot was thinking. He turned right into you!"

Horizontal on the tarmac, unsure of my injury status or ability to continue with my Big Year, I glimpsed the offending SUV down the cross street.

Wait a sec—is it still moving?

I squinted for a better view.

It is, goddamn it! It is! OMFG—this guy is going to split after running me down!

My fury overran my disorientation, and an unlikely recourse took shape. Extending a twitching finger in the getaway direction and speaking through hyperventilated breaths, I addressed the motorcyclist. "Follow that jackass and get the license plate!"

Immediately compliant, the rider scrambled back to his hog and peeled through the intersection in pursuit of the perpetrator.

Still pinned to the blacktop, I took a few deep breaths and flexed my appendages. When the stab of a broken bone or the throb of a torn tendon didn't materialize, I untangled myself from the bicycle and stood. My left forearm was scraped, but a subsequent full-body pat down revealed no serious pain, my rib healed by that March 20th juncture. I knew parts of me would be sore when my adrenaline subsided, but it appeared I'd escaped the crash without injury.

Two of my panniers had detached, so I retrieved the pair and toted them onto the sidewalk. Gawking motorists offered zero assistance, and I confronted resumed traffic when I pivoted to retrieve the bicycle. One self-absorbed jerk honked at me—presumably to hurry—as I limped my transport out of the intersection. Dumbfounded by the lack of empathy, I inspected my bicycle and the contents of my panniers for damage while I waited for the motorcyclist to return with the offender's details.

He reappeared ten minutes later. We convened in an adjacent parking lot, and he explained that he had confronted the driver after she parked in her driveway at a nearby housing development.

He continued, "She was getting her two toddlers out of the car, and I told her she needed to go back to Route 200 to check on the cyclist she'd hit. She told me I was 'full of shit' and to 'get the hell off her property' before she called the police."

After she slammed the door, he recorded her address and license plate and returned to detail the encounter to me. I thanked him for his help and insisted he go about his business.

I was furious the driver had denied her responsibility. I wanted to ride to her house and smash her windshield, but I realized it would be difficult to make a clean getaway on a bike. Absent any other recourse, I called the police.

I explained the situation to the responder, but he was unimpressed I had the offender's license plate and address. He instructed me to sit tight while he dispatched someone, apparently a low-priority task because it would require an hour.

That incensed me. Had the woman hit another car and sped away, the police would have appeared instantly, at the scene of the accident and at her home. The lack of alacrity was maddening, but I propped myself against a telephone pole and waited for the police to arrive.

Fifteen minutes elapsed before a man in his thirties appeared from the office park adjacent to the intersection. He approached and engaged me. "Is that your bike?" he asked while motioning toward my transport.

The question felt rhetorical given I was standing two feet from it, so I replied with attitude. "Yeah. So what?"

"Did you just get hit by a car?"

I couldn't contain my frustration. "No. I just decided to fall in the middle of a six-lane intersection," I said.

Deflecting my sarcasm, he said, "The person who hit you is my wife. But she didn't realize what happened. When she got home, some dude on a motorcycle showed up and accused her of injuring a cyclist. She got scared, went inside, and called me to investigate since I work nearby."

I realized his intentions were pure and apologized for being snarky. He understood my frustration and called his wife, who immediately returned to the scene with two bawling toddlers strapped into their car seats. She apologized and claimed she didn't hear the strike, a defense I believed with Thing One and Thing Two likely screaming bloody murder in the backseat. The couple agreed to stay until the police arrived, and we made small talk while we waited for the authorities.

A squad car arrived twenty minutes later, and I explained I didn't want to press charges given Susan's return and contrition. The officers were glad we'd reached an understanding and had us complete some paperwork before we went our separate ways. I rode the two remaining miles to my motel and collapsed onto the bed, the episode a sobering reminder of the danger I'd face for another nine months.

That was the first time a vehicle had struck me, but I'd been nearly flattened on several occasions. Just a week prior, a driver tore down his driveway as I approached it on the residential street. He had an unobstructed view of me, and the law dictated he yield to me before turning onto the public thoroughfare. When it became apparent he didn't see me or wouldn't yield my right of way, I yelled through his open window. That brought him to attention, and he skidded to a stop ten feet short of broadsiding me, his coffee drenching him during the deceleration. I should have slowed as a precautionary measure, but I was sick and tired of being disregarded when I had the right of way.

From the moment I conceived the trip, I understood the risks to which a bicycle Big Year would expose me. Nearly 1,000 cyclists are killed by motorists in the United States each year, and over 100,000 more are severely hurt or permanently injured. The US is a nation of cars, and most drivers don't possess the experience or patience to check for cyclists each time they turn or merge. Motorists are comparatively protected, and a collision that scratches a car can mangle a bicycle and kill a rider. Worse—and as I'd experienced in the Carolinas and Georgia—some drivers enjoy harassing cyclists; no amount of caution can protect a rider from those deranged imbeciles. When I considered every pitfall, the odds of surviving a year on a bicycle were daunting.

Conceding that I could exercise every precaution and still end up crippled or dead was an important prerequisite for my undertaking. I had to acknowledge the risks but subsequently forget about them; if I dwelled on negative outcomes, each day would become a futile exercise in trying to predict the unpredictable. I had to manage my fears, not surrender to them; all I could do was evaluate each moment and react with survival in mind.

That read-and-respond attitude was a remarkable turnaround from my postdoctoral fellowship. I'd exercised an obsessive need for control across those years, and I'd exhausted my intellectual patience trying to bend the biology to meet my expectations. Unlike an experiment where one premeditated parameter is varied at a time, the road is an open system. I couldn't control drivers any more than I could the weather, and I knew a stinging insect or overlooked pothole could send me to the tarmac at any time. It was, however, in accepting, even embracing, such uncertainty that I'd grown as a person. Flat tires had frustrated me and falls had fractured me, but my bicycle experience was formative specifically because I pushed through whatever insults and injuries my journey had supplied.

I thought more about risk as I powered into and through the Florida Panhandle over the next week. By the time I reached Pensacola, I understood my biggest risk wasn't on America's roads; it was in Boston. Had I stayed in my laboratory, I would have risked putting more time into a pursuit that I didn't love anymore. I'd risked allowing my insecurities to govern my thinking and, ultimately, making myself miserable. Worse, I risked denying myself the window the bicycle afforded, my past and present resolving with each mile I pedaled. I had no idea to what, if any, understanding my journey would deliver me, but I was proud to have boarded the bicycle against fears and uncertainties that would have dissuaded most. Trusting broken bones would heal faster than broken dreams, I powered into Alabama.

TEN

Lonely Dog

I rounded Mobile Bay, Alabama, on March 31st and continued south toward Dauphin Island, a beachfront destination featuring some of Alabama's best birding. While I was stopped at a traffic light short of the elevated causeway leading to the barrier beach, another cyclist approached and engaged me. Tall, lean, and in his late fifties, he introduced himself as Ralph, said he lived on Dauphin, and asked about my travels. I offered my stock spiel and explained I'd arrived at the island en route to Mississippi, Louisiana, and Texas. A cross-country cyclist in his younger years, Ralph fondly recalled similar wanderings. He was an enthusiastic and entertaining distraction, and the next fifteen minutes evaporated as we yakked about life on two wheels.

"You have a place to stay tonight?" he asked.

I replied, "Yeah, I've prepaid for a room at the motel."

He pressed, "What about tomorrow?"

I answered, "Dunno. If the birding is good, then I'll stay. It if sucks, then I'll make miles west."

"Makes sense. Take down my phone number anyway. If you decide you want to stay tomorrow night, I got a bed for ya. Call or swing by. Whatever," Ralph said casually, as though inviting an old friend over to watch a football game.

I appreciated his offer and input his details into my phone before we parted ways, I over the causeway and he along the mainland.

I spent that afternoon and the following morning, April 1st, exploring Dauphin. Shorebirds probed the beaches, and a smattering of songbirds presented in adjacent thickets; the additions of Whimbrel, Gull-billed Tern, Red-eyed Vireo, and Yellow-billed Cuckoo pushed my total to 270 species. Featuring a warm Gulf breeze and no mosquitos, the birding was idyllic. I wasn't ready to pull the plug on paradise by the time I sat down for lunch, so I called Ralph to inquire about extending my stay.

"Swing by any time. I'm around. Looking forward to it," he said.

I birded through the afternoon, grabbed dinner, and headed for his place as sunset approached. Standard-sized and raised by two feet of brick skirting, Ralph's mobile home looked welcoming enough. Two rutted tire tracks led from the street to a shed-type garage. A lawn populated by a flock of decaying plastic flamingos exuded a unique, bayou-beach brand of chic. I leaned my bicycle against the front hand-rail and rang the doorbell.

A deafening cacophony of barking, baying, and meowing erupted inside the house; accompanied by a whirlwind of scampering, it was bed-lam. Images of seething, blood-thirsty beasts flashed through my head, and I wondered how many animals were required to create such chaos. *Five? Ten? Fifteen?*

Paralyzed by curiosity, I watched the door swing open. Furry faces were everywhere: on the sofa, behind the recliner, under the table, and in the sink. They peeked around corners and emerged from shadowy recesses. No matter where my gaze wandered, I saw snouts, paws, and tails. I thought it was the best April Fool's joke in history until the smell of stale urine struck me, two excited dogs actively contributing to the overpowering odor as I tried to comprehend my predicament. I wasn't sure if staying was advisable, but Ralph put his arm around my shoulder before I could evaluate a retreat.

"This way, c'mon!" he said. "I knew you were an animal lover. I could feel it on the road yesterday!"

He led me through the small living area and down a short hallway to a tiny guest room. I crossed the hallway to the bathroom, shooed a fluffy feline off the toilet, booted a medium-size mutt out of the tub, and took a shower.

Refreshed and ready to face the onslaught a second time, I joined Ralph in the living room. He was in his recliner with two dogs in his lap, and I repositioned several others to create space on what remained of the decrepit couch. I noticed an Auburn Football banner on the wall, and we initiated a lively discussion about the recently implemented College Football Playoff before rejoining our shared interest in cycling. The animals were squabbling, barking, meowing, and peeing throughout, but our conversation flowed as freely as urine despite those distractions.

Thirty minutes into our exchange, a car pulled into the driveway. Ralph popped out of his recliner and exclaimed, "That's my wife!"

The four-legged frenzy grew in anticipation of her entrance. Ralph introduced me to Mary and explained I'd be spending the night. I thought it was strange that he hadn't informed her of my stay, but she seemed fine with the idea.

"Oh well, what's one more? Welcome, dear. Anything you need, you let me know," she said with a smile.

Mary's tone echoed Ralph's, and I immediately understood their attraction. The commotion surrounding her arrival dissipated, and the three of us settled in for an evening of television and conversation. Squeezed onto the couch with the animals and me, Mary was charming. Everything was in perfect loving order save for the attending animals. I didn't want to ask the obvious and intrusive question about their herd, but the conversation slowly tacked in that direction. The couple volunteered their story, and I sat rapt as it unfolded.

Ralph began, "It started five years ago. We had recently adopted a dog, and I found another on the side of the road. He looked like he needed a friend, so I brought him home."

Mary chimed in, "And I always wanted a cat, so we got one from a local shelter. But then I found another one in the yard and adopted it, too."

Ralph continued, "I found another stray dog, and then another, and then another, and I brought them all home since they had nowhere else to go. Soon, people started tying unwanted dogs and cats to our front railing in the middle of the night."

"They all needed our help, and we couldn't turn any of them away," Mary added.

It was obvious that the couple never intended to have thirty pets, but their bleeding hearts guaranteed that outcome once they'd adopted the first. My experience with alcohol suggested there was an underlying psychology to motivate such binge behavior, but I held no desire to identify it. Emotional dissection is most effective when initiated on one's own timetable, and their assembly looked healthy and happy in the meantime. Their intention was pure, and I was thankful Ralph had invited me to stay the night. Remembering that I had a long ride the following day, I excused myself and retired, the gentle pitter-patter of paws as soothing as soft rain while I nodded off.

The next morning, I bid Ralph and Mary an early goodbye and cranked onto the causeway as the sun cracked the horizon. Golden rays danced across the Gulf's surface, and I drew deep breaths of salty air to flush the lingering animal dander from my lungs.

Exiting the span and rolling through the intersection where I met Ralph two days prior, I finally made sense of our interaction. His lodging invitation seemed forward at our introduction but made perfect sense after our time together; to Ralph, I was just another lonely dog looking for a warm place to sleep. His boundless empathy guaranteed that he couldn't help himself from helping me, and I would have denied us both had I succumbed to reflex and run at the first, gut-wrenching whiff.

Ralph exemplified the best of my journey to that point: the friendly people. I'd figured out what to expect from the birds and the bicycle, but the folks I met along the way were a treasure trove of personalities, quirks, stories, and support. For every dumbass who harassed me on the road, five curious folks offered to buy me lunch and a dozen generous souls opened their homes to me. The unyielding assistance meant the world to me, and I hoped that hearing about my adventure inspired a few folks to dream a little larger. Steering through the shrimping mecca of Bayou La Batre, I headed for Mississippi.

ELEVEN

Creating Community

Hurricane Katrina made landfall as a Category Three storm in Southeast Louisiana on August 29, 2005. With sustained winds in excess of 120 miles per hour, the system caused $120 billion in damage. The images from New Orleans were heartbreaking; the levee failures displaced tens of thousands of residents and killed over a thousand people. Despite those horrors, the city was spared Katrina's strongest push. That furious honor, a result of more intense surge to the east of the storm's eye, was reserved for Coastal Mississippi.

Like most around the country, I watched the coverage of Katrina's devastation but succumbed to shinier headlines in subsequent weeks. Exploring Pascagoula, Mississippi, on April 2nd, 2014, I realized that was a mistake; my temporary diversion was an ongoing, nine-year ordeal for those who experienced the storm. Exposed foundation slabs lined the beachfront, and the surviving or since-built structures loomed arachnid-like atop exaggerated stilts. Branchless oaks haunted Highway 90 as I moved west through Biloxi, and I felt as if the pause button had been pushed on the entire Mississippi Coast. In Gulfport, vacancy had overrun vibrance, and my inability to find an ice cream shop, a staple of any healthy beach community, suggested recovery was still incomplete nearly a decade after Katrina.

That afternoon, I met up with Michael Sandoz, a local birder, in Long Beach, Mississippi. Only sixteen years old, Michael had found my blog a few weeks prior and contacted me about hanging out as I passed through the area. He arrived on his bicycle, and the two of us took in views of Black-bellied Plovers, Ospreys, and Least Terns (#271) as we pedaled along the beachfront. A few minutes later, Michael suddenly dismounted his bike and ran into a thicket. I caught up with him, and he pointed out an Orange-crowned Warbler before darting off in another direction. Kooky and confident, he reminded me of myself at the same age.

I overnighted with Michael and his family, crossed the Pearl River into Louisiana, and pedaled toward New Orleans. Reaching the French Quarter after eighty-two blustery miles, I checked into a hotel, cleaned myself up, and ventured out to explore the city. It was my first visit to the Big Easy, and I soaked in the street vibes as musicians and entertainers performed on every corner. One clarinetist, a woman in her fifties, alternated between forever-held notes and dizzying displays of dexterity as she manipulated the valves and levers on her instrument. Tourists strolled while locals lolled, and a feeling of kinship washed down the same streets as Katrina's surge had. My commute into New Orleans suggested that residential neighborhoods hadn't experienced similar revitalization, particularly those that appeared minority or lower income, but life in the affluent and tourist-facing French Quarter had rebounded.

A dinner of boudin and seafood jambalaya was a welcome break from fast food, and I walked back to my hotel with a full stomach and burning lips sometime after nine o'clock. Sleep should have been easy after the day's long ride, but I was restless past eleven. Street sounds sowing curiosity, I redressed for a midnight stroll.

Bourbon Street was alive at that hour. Neon signs competed for the attention of glassy-eyed partygoers, and plastic beads crunched underfoot as groups of interlocked friends staggered between watering holes. With music escaping bars and balconies, disinhibited throngs sang and danced as they flowed up and down the street. Women batted eyelashes, men leaned in for kisses, and smiles and laughter rendered worries forgotten. I was a sober ball in a drunken pinball machine, and I thought it was equal blessing and disappointment that I didn't experience New Orleans during my drinking and drugging days.

I was 22 years old when I graduated from Stanford and moved to Cambridge, Massachusetts. I found a basement apartment two blocks from Harvard Yard and settled into the Department of Molecular and Cellular Biology as a research assistant for Elizabeth Robertson. She was internationally recognized for her work with embryonic stem cells, and her group, a brilliant assembly from around the globe, studied the genetic specification and morphological organization of tissue layers—gut, muscle, nervous, and so on—in the developing embryo. I was deployed as experimental support for graduate students and postdoctoral fellows, and I hoped my time in her lab would grow my experimental capacity before I applied to PhD programs in two or three years.

My labmates were a fun bunch, and we frequented Shay's, Charlie's Kitchen, The Hong Kong, and other Harvard Square institutions for afterwork drinks. I always consumed three times as much as anyone else, and I invariably extended my intoxication after they retired. A social chameleon, I was comfortable in any setting and didn't hesitate to venture to new places on my own. With my DJ interest lending connection to other nightcrawlers, it didn't take me long to infiltrate the music scenes in Boston and Cambridge.

That process accelerated in November, while I was shopping at Satellite Records, a purveyor of underground dance music. Scanning my vinyls, the twenty-something clerk spoke up. "You've been in here a lot the last few weeks. You new in town?"

My experience suggested that record-store employees operate in the space between aloof and condescending, so his unsolicited outreach was surprising. I replied, "Yeah, man. Just moved here from California. Slowly figuring it out. Lotta nights in Central Square since it's close to home. Phoenix Landing, Man Ray, Enormous Room. Liking it so far."

"Well, if you like those places and you're buying this music, then you gotta check out Rise." He slid something across the counter and continued, "I'm spinning there on Friday. You should swing through."

I picked up the slick purple business card. Examining the logo and address, I said, "I've heard whispers about this place but don't know much about it."

"Best place in Boston. Wicked music, cool peeps, no bullshit," he said. He explained that the club was members-only and open from one thirty a.m. to six thirty a.m. on Saturday, Sunday, and Monday mornings. The guest pass plus $20 would gain me entrance.

"I thought everything shut down at two a.m.?" I asked.

"Not Rise. They don't sell booze so they can stay open all night. Lotta other stuff going on though."

I understood his insinuation. "Sounds fun," I replied. "I'll try to make it."

A friend from high school had invited me to a house party at her place on Friday, so I spent four hours there, chugging beers and downing shots, before hooking a cab toward 306 Stuart Street at two a.m. Despite arriving alone and shit-faced, I was granted admission to the nondescript building after the doorman scanned the barcode on the backside of the guest pass.

Climbing a narrow wooden staircase, I found myself in a plush lounge furnished with oversized couches. Deep house grooves and soft purple lights set a mellow and interactive vibe, and the amount of close-talking and arm-stroking suggested that most were afloat on ecstasy. My curiosity piqued, I climbed a second flight of stairs, opened a metal door, and walked into a converted attic. Mirrors covered the walls, and an array of two-dozen disco balls hung from the low-slung ceiling, colored lights turning the reflective room into a gigantic kaleidoscope as they swept across it. With a hundred bodies pulsating to melodic synths and tribal beats, it was liked I'd stumbled into heaven.

I began introducing myself, and the hours evaporated as I rubbed elbows with the regulars, some of whom I recognized from other venues around town. I learned that the club relied on annual membership dues and that each member is allotted a small number of guest passes each year, an identifier on each pass allowing the club to know who invited whom. One tall and lanky guy, Dave, furnished me with another pass so I could return the next week.

"Keep coming back. Once you meet a couple more people, you'll be proposed for membership," he said. "It's no biggie. We just want to know you're not an asshole."

A month and three visits later, Dave proposed me as a Rise member. Two months after that induction, the owners made me an official Promoter, which granted me unusual latitude to recruit new members from around the city. I was honored the club trusted me to bring whomever I wanted into the fold.

Rise was more than a drug-fueled, after-hours playground; it was a community of people brought together by their love of dance music and the positive world view—peace, love, unity, respect—which it engendered. Different races, nationalities, and sexual orientations were represented, and judgment was universally left at the door, a circumstance that guaranteed no one questioned the drinking I did elsewhere or the drugs I did once I arrived. Cocaine inevitably inserted itself and, alongside ecstasy, kept me going at all-morning after-parties that started when the club closed at six a.m. I had endless energy, so I always recovered by Monday. A couple of nights of blackout drinking at bars and lounges broke up the work week, and by Friday I was ready for another turn or two at Rise.

Toward the end of my first Boston year, I drove to Hotchkiss to attend my fifth-year reunion, a three-day, on-campus event held alongside the tenth-, fifteenth-, twentieth-, and thirtieth-year iterations. Most attendees enjoyed an afternoon beer and a few evening cocktails as provided by the school, but I brought supplementary alcohol and was annihilated the entire time, including the morning round of golf I played with a few buddies. It was great to catch up with friends and faculty, but several people asked if I was OK, their insinuation being that I had a drinking problem.

In each instance I replied, "I'm good, thanks. Killin' it in the lab. Grad school is going to put a lid on this in another year or two, so I'm going to have fun until then."

It was the same flawed rationale I'd offered Brynn, my Stanford dormmate, four years earlier.

While my classmates let it go, the campus librarian pressed me harder. I didn't have a lot of day-to-day interaction with him while I was at Hotchkiss, so his outreach was unexpected.

"You are one of the most intelligent and charismatic people who's gone through here, but you're doing yourself a disservice with the drinking. You think you're fine, but you don't have the age or experience to know what you've gotten yourself into."

My relationships with my Hotchkiss classmates slipped in the ensuing years, but I was looking ahead rather than behind. I was incredibly productive in my Harvard laboratory through my second Boston year, and I'd secured authorship on several papers by the time I applied to graduate school in the fall of 2003. Despite what I'd articulated to others, I was curious if that next intellectual step would govern or grow my behavior. In sober moments, I hoped academic demands would quash habits that I knew wouldn't serve me in the future; in the intoxicated next instant, I abused the flexibility that academia offered in order to accelerate my condition. But that's one of alcoholism's fortes; it creates doubt and casts itself as the only answer.

My eventual choice pitted the University of California at San Diego and the Salk Institute against New York University, both with outstanding programs in developmental biology. I was unable to draw scientific distinction between the two institutions, so my affliction assumed the responsibility for me. With a drunk-driving conviction from my time at Stanford, I understood New York's subway would tolerate behavior San Diego's freeways would not. Framed that way, my choice was easy: move to a city where I never had to drive. Arriving at NYU in the fall of 2004, any hope that graduate school responsibilities would temper my behavior was gone. I was totally fucked up—even when I wasn't drunk.

———

Part of my program since getting sober was avoiding settings like Bourbon Street, but I was too curious to let the spectacle go unexplored. Watching people stagger along, I was envious of their incapacity. I missed alcohol's numbness, ecstasy's euphoria, cocaine's confidence, and the tales these Bourbon revelers would tomorrow have to tell. The hardest part of my sobriety was routine, and I'd sometimes struggled to find excitement without the social, sexual, and pharmacological possibilities that alcohol and drugs supplied. Gazing at Bourbon Street, I missed drinking and drugging, but not enough to reenlist. Sobriety would only be as predictable as I allowed it, and I trusted my days on the bicycle would synergize in ways that my drunken nights didn't. My curiosity satisfied, I returned to my hotel and nodded-off.

I spent the following day exploring City Park before exiting New Orleans on April 5th. Thick fog brought about several wrong turns as I fumbled toward the Huey P. Long Bridge, and my satisfaction from crossing the Mississippi River evaporated on the span's exit ramp when I heard an audible pop. My rear tire went flat the next instant, and I struggled to control my fishtailing transport against the momentum I'd gathered on the descent. A grassy margin appeared, and I piloted my wounded bicycle toward it before collapsing onto that cushion.

Calmed and collected after the brief scare, I took the wheel apart, removed the offending three-inch nail, swapped in a new tube, and continued after a fifteen-minute delay, my repair skills honed since my Bayonne baptism.

The bicycle felt increasingly sluggish across the next five miles, until I realized the back tire was slowly losing air. Frustrated by another delay and unable to find whatever caused the slow-leak puncture, I hastily swapped in my last replacement tube, that carelessness to blame when I suffered a third flat twenty minutes later. I was furious at myself for overlooking the tiny glass shard before it cut two tubes, but I salvaged one of those with my final patch. The wheel and bicycle reassembled, I resumed my course.

Worried that my tires and tubes wouldn't hold until I reached Morgan City, I missed a turn at Raceland, continued west on Highway 90, and rolled onto an elevated causeway where traffic assumed interstate character and velocity. The skinny shoulder was littered with fast food trash, broken bottles, and car parts, but I continued nervously forward because I assumed the configuration was temporary; only after five harrowing miles did I consult the map and realize I was crossing the Atchafalaya Basin, the largest cypress swamp on the continent.

The eastbound and westbound lanes occupied independent causeways, so I decided to continue rather than backtracking against oncoming traffic. Louisiana's roads were the worst I'd suffered, glass seemingly sprouting out of the tarmac, and my tire predicament suggested I should take the most direct route across swamp lest I suffer another flat without the supplies to repair it.

If continuing wasn't technically illegal, then it was highly discouraged by eighteen-wheelers thundering by my left ear at seventy miles per hour. I doubted that I'd see the inside of a holding cell if law enforcement intervened, but I'd certainly be escorted off the span in a police cruiser, a result which would shatter my precious, petroleum-free streak. Prior to departing Massachusetts, I vowed the only ride I'd accept would be in an ambulance. Encountering Barred Owl roadkill every other mile, that species common in cypress swamps, I felt the weight of that possibility as I raced across the Atchafalaya.

The transit lasted an hour, and I reached Morgan City without energy or adrenaline to spare. Dragging my bicycle into my motel room, I noticed my rear tire was losing air for the dreaded fourth time, my final tube punctured by another glass shard, presumably right at the end of my sprint. Inspection of the overlying tires revealed them shredded; it was a miracle they'd held. The local Walmart would stock patch kits to repair my punctured tubes, but undertaking that eight-mile round-trip walk without also replacing the tires was pointless. With nowhere to buy those specialty items in the tiny town, I'd need to pause for at least two days to have new tires delivered.

Thirty minutes after blogging about my predicament, I received a phone call from Crystal, a blog reader in Baton Rouge. She suggested I call her local bike shop the following morning and have them round up the required supplies. I could pay for them over the phone with my credit card, and she'd collect and deliver everything to me by noon, a brilliant plan that would leave the afternoon to ride the fifty miles to New Iberia, where I'd arranged an overnight with a local cyclist. I immediately accepted her offer and slept well ahead of our rendezvous.

Crystal was a professor of environmental science at Louisiana State University. Displaying rich caramel skin and rocking a beaming smile, she was friendly and talkative at our in-person introduction. She was particularly curious about my academic departure, a decision I'd referenced in multiple blog entries. She explained, "I love science, but it's so tedious and uncertain. You can do years of experiments and have nothing to show for it."

I empathized. "That was my postdoctoral fellowship. I hit an insurmountable roadblock and had to decide between restarting with another project or getting out. I had no idea what life outside the lab looked like, so I cooked up this ridiculous idea to give me time to figure it out."

"I just can't believe you're doing it. Research, teaching, and grant writing pull me in so many directions, but I'm too tired, too scared to see what else is out there. That's why I enjoy reading your blog. It's a total escape."

I didn't think Crystal was ready to jump ship, but it was amazing to see how her thinking mirrored mine. She'd ascended higher on the academic ladder, so it felt odd to dispense career advice to someone my senior.

She eventually asked the inevitable question, "What do you think you'll do when the year is over?"

Still working to understand my past, I hadn't turned an eye to the future. I knew I'd need to face it at some point, but I wasn't ready to climb that mountain from the Louisiana swamps. I answered, "I'm tryin' not to worry about it. Too much focus on the future ruins the present, ya know? But I'm kinda hoping I haven't imagined the outcome yet. I just gotta keep pedaling and see where I end up."

Rolling along on new, puncture-limiting Kevlar tires, I thought about how my blog had brought Crystal and me together. Enjoyed by birders, cyclists, nature enthusiasts, and environmentalists, the online diary was a public portal that agglomerated all sorts. Comments allowed readers to communicate with me and interact with each other, and I was proud to have created a forum where far-flung people could convene for a few vicarious minutes each day. Cruising west through crawfish ponds and rice fields, I realized community could as easily be built on biking, birding, and blogging as drinking, drugging, and dancing. No one in sight, I felt more connected than ever.

TWELVE

A Story of Wind

My sprint across the Atchafalaya Swamp had highlighted a critical consideration for long-distance cyclists: the trade-off between safety and expediency. There were alternatives to the Highway 90 causeway after I inadvertently committed to it, but those would have added distance, hastened the demise of my tubes and tires, and likely stopped me short of Morgan City, a destination I barely reached as it was. Similarly, I could have avoided Highway 17 through the Carolinas and Georgia, but back roads would have multiplied my miles and slowed my progress. The Big Year clock was ticking, and I needed to stay on pace if I wanted to keep the possibility of 600 species in play. My total was at 277 when Crystal came to my rescue on April 7th.

Birds also balance safety against speed, but their outcomes are shaped by evolutionary influences rather than conscious consideration. Many neotropical migrants—species that nest in the United States and Canada and winter in Latin America—return north through Northern Mexico because that overland route offers consistent opportunities for food and shelter. Others forgo that safer option and instead cross the Gulf of Mexico, the distance from the Yucatán Peninsula to the Gulf Coast overcome as a 600-mile, nonstop flight. The trans-Gulf option is faster but can turn deadly because the migrants can't rest or refuel over the ocean.

How trans-Gulf migrants fare on the crossing is a story of wind. When high pressure prevails over the Gulf, the prevailing breezes blow from the southeast. Under such favorable conditions, the trans-Gulf migrants depart the Yucatán at sunset, ride the tailwinds north through the night, and reach the Gulf Coast sixteen to twenty hours later, between ten a.m. and two p.m. the following day. Among the cuckoos, thrushes, flycatchers, tanagers, warblers, vireos, and orioles that undertake the fantastic flight, the tiny Ruby-throated Hummingbird is particularly impressive; weighing less than a nickel, the diminutive dynamo completes the overwater haul exactly as the larger birds do.

While trans-Gulf migrants have evolved the ability to assay the wind direction at their Yucatán departure point, with northern breezes (headwinds) dictating a delay to a subsequent night, the birds cannot anticipate a shift that occurs mid-crossing. That scenario unfolds when storms push southeast through the Great Plains, onto the Texas and Louisiana Coasts, and over the Gulf overnight or in the early morning hours. The migrants tire when they encounter soaking rains and impeding headwinds, and some percentage suffer terminal exhaustion and plunge to watery graves. The strongest storms can wipe out a significant portion of a night's flight, but heavy losses are rare and statistically overcome by the aggregate number of migrants that survive the crossing on favorable nights. While trans-Gulf migration is a huge gamble for individual birds, it has evolved because it delivers the surviving population to the breeding grounds quickly and efficiently. On bird wings or bicycle wheels, the fastest route is rarely the safest.

Among many places to intersect the arriving trans-Gulf migrants, the Texas coast—specifically the stretch from Galveston east to the Louisiana state line—is arguably the best. Migration peaks in the second half of April, so I hoped to reach the region on 15th and remain through the 30th before continuing west toward Arizona. A timely arrival was critical because virtually all of the trans-Gulf migrants breed in the eastern half of the continent. If I missed the northbound birds as they passed through Texas, then I'd render my chances at 600 species as flat as my tires in Morgan City. I planned to return to Texas in mid-November, but the migrants would have left their breeding grounds and retired to the tropics by that juncture. It was therefore "spring or bust" on the Texas coast.

Given my cycling inexperience at adventure's outset, atrocious winter weather in the Northeast, aggressive drivers in the South, the vehicular strike in Florida, and tire-shredding roads in Louisiana, my timely arrival in Texas on April 12th was nothing short of a miracle. I'd covered 4,500 miles since departing Salisbury, Massachusetts, and reached Texas three days ahead of my most-optimistic projection. I experienced a huge sense of accomplishment, and I basked in the confidence that the next two weeks would feature maximum birding and minimal biking.

Sadly, Sabine Pass was less inspiring. Positioned on the Texas–Louisiana border, where the Sabine River reaches the Gulf, the town of 2,000 had been decimated by hurricanes twice in the previous decade, first by Rita in 2005 and later by Ike in 2008. The place was a ghost town six years after that second strike, and my threadbare motel and the adjacent Chevron station were the only brick-and-mortar businesses in evidence, the half-vacant municipality mostly a crash pad for petroleum industry transients, as far as I could discern. Foraging beyond the gas station was futile; dinner that night—like the following four—was a soggy Hunt Brothers pizza, a can of room-temperature peas, and a pint of Blue Bell ice cream. With breakfasts and lunches to come from that same culinary black hole, escaping Sabine Pass before I succumbed to congestive heart failure or constipation would be a challenge.

I woke early on April 13th and pedaled five leisurely miles to Sabine Woods, a thirty-acre reserve positioned on the beachfront. Exploring a meandering network of trails on foot, I spotted a rufous Wood Thrush stalking prey in the understory and a pink Summer Tanager plucking bees from a hive. Bird activity was low, but I was happy to spend the time strolling instead of cycling.

Rounding a tight bend in the path, I nearly collided with an excited preteen boy. His binoculars suggested he was a birder.

"I'm very sorry, sir," he said politely.

"Sir? Let's ratchet the formality down a bit, kiddo. Whatuvya found so far?" I asked.

He spoke confidently, "A Kentucky Warbler next to the creek and a Yellow-billed Cuckoo near the back."

"You know your birds. What's your name?" I asked.

"Chris. And Grandma Lily is here somewhere," he replied.

Perfectly cued, a white-haired woman hobbled around the bend and spoke through labored breaths. "He moves too fast. I can't keep up!"

They were from Houston and, like me, making their first visit to Sabine Woods. Chris commented on the lack of activity. "I thought there'd be more birds, like the pictures on the internet. What's going on?"

I answered, "Well, the migrants that arrived yesterday continued north last night, but the next round hasn't arrived yet. They usually get here around noon, but we won't see most of them."

I explained how the southeastern tailwinds offer the migrants such huge assistance that most don't need to rest upon reaching the Texas coast; instead, they stay in their high-altitude migration lanes and wing deeper into the continental landmass, the tailwinds offering a veritable free ride north. "For every bird we see in the reserve today, a hundred more will have flown right over us, some a mile high."

Chris wanted to know more. "So, when does it get really good, like when birds are everywhere?" he asked.

"That happens when a storm comes through, usually from the northwest. The birds don't want to fly into the wind and rain, so they stop at the coast to rest," I explained.

The biggest events, referred to as "fallouts," occur when migrants are forced to battle inclement overwater weather for hours. Exhausted survivors collapse onto the first ground they encounter, and the catatonic specimens allow birders very close approach. The views are amazing, but it's depressing to know many of their travel companions perished under such stormy circumstances.

I continued, "Sun and southeast winds are good for migrants because they make the crossing easy. The birds aren't tired and most overfly the coast. Rain and north winds are good for birdwatchers because the birds are tired and forced to land when they arrive. Does that make sense?"

Chris nodded and said, "It's going to rain tomorrow, so it should be good birding in the afternoon, right?

"Yes, assuming the front doesn't peter out before it reaches the coast," I answered.

Chris pleaded, "Grandma Lily, can we come back tomorrow if it rains?"

She wisely passed the buck. "It's a school day, so it'll be up to mom and dad," she said.

"Don't worry about tomorrow, Chris," I said. "There are some cool birds here today, so keep looking!"

My encouragement registered, and Chris tore down the path, presumably in pursuit of a bird he'd spotted. Gramma Lily thanked me for the time and headed off after him, albeit at a slower pace. I heeded my own advice and pounded the trails for the remainder of the day. Among the few new birds I found, American Redstart, Tennessee Warbler, and Northern Waterthrush pushed my total to 299 species.

I returned to my motel late in the afternoon and ate my second round of pizza and peas while watching the local news. The forecast suggested the front would hit the coast around noon, a perfect scenario. Unlike an overnight or morning storm, which would befall the migrants while they were still over water, midday rain would allow the birds to complete the crossing but force them to the ground upon arrival. With visions of vireos and warblers racing through my head, I could hardly sleep.

Southeast winds persisted when I returned to Sabine Woods the next morning, but the breezes abated in the ensuing hours and died at noon, exactly as predicted. The first flash of lightning was an encouraging sign, and I quivered with anticipation as a thunderous rumble shook the beachfront oaks. A phalanx of dark clouds advanced from the north, and rain promptly engulfed my position. The meteorological pieces in place, I peered into the tumultuous heavens.

I spotted a spec at the limits of my vision. My binoculars revealed a small songbird, and I glimpsed hints of black, white, yellow, and orange as the delicate form plunged toward Earth. A male Blackburnian Warbler resolved, and his fiery throat glowed as he closed the distance between us. Fast approaching the treetops, he broke his free fall with an acrobatic wing flip and lit on a moss-covered branch. He promptly caught and gobbled a large insect, fuel to replace what he'd burned on his overnight flight.

Behind the Blackburnian were countless beautiful others. Scarlet Tanagers and Indigo Buntings fell from the sky like iridescent raindrops, and Yellow Warblers and Rose-breasted Grosbeaks decorated trees like ripe fruits. Flycatchers swirled between perches, orioles darted between trees, and hummingbirds buzzed in every direction. The arrivals looked healthy, and I doubted the storm put any of their companions into the abyss. I marveled at their ability and endurance; even aided by similar tailwinds, I'd require a week to cycle 600 miles.

The rain dissipated after an hour, but the northern airflow persisted through the night and pinned the newly arrived migrants on the coast for a second day. The birding on the 15th rivaled the day before, and I pushed my total to 315 with the 16 new species I found across the two days. Most notable was a male Black-throated Blue Warbler. The species winters in the Caribbean and returns north through Florida, so I didn't expect it in Texas. If the storm hadn't grounded that anomalous individual west of its traditional migration lane, then the species would not have joined my year list. I didn't know at the time, but that front would prove the only weather event of the entire spring. Had Crystal not delivered tires when she did, I would have arrived a few days later and missed it.

Though it brought bountiful birding, the front didn't deliver every trans-Gulf migrant because species stagger their migrations relative to their breeding latitudes. Those nesting in the southern United States return first, while those nesting in Canada return last, the northern boreal areas requiring additional weeks to thaw. The storm conveyed all of the early birds, but I'd need to hang around the Texas coast for another two weeks, through April 30th, to find the later comers.

I spent two additional days at Sabine Woods before departing the beachfront and moving to the marshes of Anahuac National Wildlife Refuge. Four days there netted me Swainson's Hawk, Northern Bob-white, Buff-breasted Sandpiper, Franklin's Gull, and a vagrant Ruff, an Old-World sandpiper that makes sporadic cameos in the New World. That bonus added, I left Anahuac midday on the 21st and rejoined the beachfront.

I spent the next six days exploring High Island and the Bolivar Peninsula. Each of those featured sunny skies, strong southeast winds, and slow birding, but I was able to grind out representatives of later-migrating species like Philadelphia Vireo, Veery, Magnolia

Warbler, Blackpoll Warbler, and Bay-breasted Warbler by walking the same wooded trails many times over. Aided by the additions of Black Tern and White-rumped Sandpiper in marshy areas, my list increased to 332 species on April 26th.

That night I faced a difficult strategic decision because I'd found all the expected migrants except for three: Black-billed Cuckoo, Mourning Warbler, and Yellow-bellied Flycatcher. The cuckoo is hit-and-miss on the Texas coast, and the warbler and flycatcher wouldn't peak for another week, those northern nesters being among the last to migrate. I still had four days before my May 1st departure deadline, but forecasted southeast winds promised to extend the slow birding.

Against that meteorology, I considered departing the coast prematurely and continuing west with four days of temporal insurance in my pocket. I wasn't keen to leave birds on the table, the Texas coast being my last opportunity at the unaccounted trio, but the thought of investing the days and failing to find them was repugnant. I realized I could ante one day at a time and reevaluate based on what I found, but I didn't want to subject myself to iterated waffling. I wanted to be decisive, so I felt I needed to choose between exiting the next morning, April 27th, or committing to the coast through the 30th—unless I found all three birds before then.

Looking west, I had a thousand miles of riding to reach southeastern Arizona. Most of that distance would be made through scorching terrain, and logistics would assume elevated importance because services (food, water, lodging) would be few and far between in the undeveloped reaches of West Texas and southern New Mexico. My cycling would become strictly station-to-station, and I'd need to know I could complete any ride before starting it because en route bailouts wouldn't present themselves as they had in populated areas. Southeastern tailwinds would yield to western headwinds once I reached Austin, and I'd face thousands of feet of elevation gain as I braved Hill Country and continued into the Chihuahuan Desert beyond it.

For the first time in my two-wheeled travels, I was intimidated. I didn't know how my body would respond to heat and elevation, and I was worried that fierce western winds would pin me for days at a time. I'd need two weeks to overcome Western Texas and New Mexico under ideal circumstances, so I decided to face the transit with four extra days in hand. Conceding the possibility of three additional species was painful, but I felt I had more control of the bicycle than the birds, given the forecast.

Thinking like a trans-Gulf migrant, I harnessed the continuing southeast winds to make 110 miles north and west around Houston on April 27th. The assistance lessened as I pushed inland toward Austin, but I powered into that layover after another two days and 154 miles. The real challenge would begin west of that gateway, but I promised to remember the migrants when I faced it; if they trusted their wings to carry them across a 600-mile abyss, then I'd bet on my legs to deliver me through 750 miles of sweltering desolation.

THIRTEEN

Bumps in the Road

I departed Austin on May 2nd, after a two-day pause for rest, resupply, and bicycle maintenance. Wheeling west, I joined Highway 290 and gained elevation into Hill County, the rolling juniper-oak hills the first significant topography I'd encountered. Limestone bedrock and poor soil dictated ranching over agriculture, and droopy-eyed roadside bovines, mechanically chewing their cud, paid me little attention as I undulated through the arid landscape. In reaching Fredericksburg, Texas, for the night, I netted 1,000 feet of vertical gain from 2,800 feet of climbing spread across 75 miles.

I continued west on Highway 290 the next morning and intersected Interstate 10 midday. Stretching 2,460 miles from Santa Monica, California, to Jacksonville, Florida, I-10 is the most direct route across the southern United States. I'd paralleled the transcontinental for several stretches but not utilized it because, besides being dangerous, cycling on interstates is illegal. Local streets and state highways had offered quieter riding and better birding without significantly increasing my daily mileage, so I didn't consider the interstate until I pedaled into the undeveloped wilds of Central Texas. There are alternatives to I-10 through some of that geography, but none feature adequate support; with interchange towns and associated services—food, water, lodging—spaced at thirty-five- to seventy-five-mile intervals, I gauged I-10 was my best chance to survive the otherwise inhospitable landscape. Interstate prohibitions are generally relaxed in undeveloped areas, where interchanges are very few and far between, but I suspected my fate hung more on an intervening office's mood than legalese.

Expediency and logistics aside, I also hoped I-10 would insulate me from the harassment I'd experienced throughout the South. Between justifying my interstate presence to an officer or escaping a truckload of hooligans on a deserted backroad—the sort that shot and killed Australian baseball player Christopher Lane "for fun" while he was jogging on a rural Oklahoma road nine months prior—I chose the former. Unless forcibly removed, my plan was to ride I-10 from Central Texas to Arizona, a distance of 650 miles, the only anticipated interruptions in El Paso, Texas, and Las Cruces, New Mexico, where surface streets would present alternatives.

Although I-10 swells to a dozen lanes through major cities, the artery presented as two-lane parallels at my intersection. A wide median separated the directional traffic, and the tarmac stretched thinly over the rolling hills before disappearing into the distance. Cars and trucks dotted the blacktop at irregular intervals, the most-distant hardly moving in the context of the enormous landscape. Unsure how my bicycle would play with transcontinental traffic, I took a deep breath and powered up the entrance ramp, onto the westbound shoulder.

My apprehension abated through the next hour. The breakdown lane was wide and well-surfaced, and a deep-cut rumble strip would alert drivers who drifted in my direction. Sweeping turns guaranteed I was always visible, litter and debris were absent, and a trickle of traffic ensured I could flag someone down in case of an emergency. The interstate seemed a wise decision, and reaching Arizona seemed straightforward at that rosy moment.

I reached Junction, Texas, midafternoon and used the remainder of the day and the following morning to explore South Llano River State Park. The mesquite-dominated habitat was the driest I'd encountered, and desert species such as Black-throated Sparrow, Canyon Towhee, and Scott's Oriole presented as I searched for Black-capped Vireo, a Central Texas prize and conservation success story.

Yellow-olive above and creamy white below, the sparrow-sized Black-capped Vireo sports white feathers around the eyes and across the lores, those spectacles imbuing the bird with tender and inquisitive visual qualities. Their population plummeted to 300-some individuals in the late 1980s because of habitat degradation at the hands of ranching and nest parasitism by Brown-headed Cowbirds, but management on both fronts has facilitated a remarkable resurgence.* By my 2014 arrival, the breeding range of Oklahoma, Texas, and Northern Mexico hosted upward of 10,000 birds. With 96 percent of the acres in Oklahoma and Texas privately held, conservation entities and government agencies liaised closely with local landowners to reduce ranching impact and implement cowbird trapping. Listening to a male vireo belt out his song, a jumbled series of buzzy whistles, I hoped the recovery would inspire future cooperation between biologists, government, and citizens.

I rejoined I-10 after securing the vireo for species #353, but my course cratered when a series of deep gouges appeared in the break-down lane. Positioned outside the standard rumble strip at thirty-foot intervals, each gouge was like an inverted speed bump and delivered a bone-shaking double jolt as my tires dropped into it.

Tha-thunk!

The gouges were unavoidable because they spanned the full width of the breakdown lane. I slowed to minimize each impact, but the forceful collisions rattled my bicycle and body. It felt like my teeth were going to shake loose, so I clenched my jaw to keep them in my mouth.

Tha-thunk!

My frustration multiplied with each blow, my joints hurting and my ass aching as I continued into the torturous configuration.

Tha-thunk! Tha-thunk!

I guessed the gouges were designed to alert unaware drivers but wondered how narcoleptic or shit-faced a motorist would need to be to ignore the aggressive vibrations induced by the standard rumble strip.

---

* Female Brown-headed Cowbirds lay their eggs in other species' nests while the host parents are away. Cowbird chicks hatch quickly, grow rapidly, and subsequently push host eggs and young out of the nest. The host parents then unknowingly raise the cowbird chicks as their own. Cowbirds prefer open country, so continent-wide deforestation has allowed the species to assert itself at the expense of other birds.

Tha-thunk! Tha-thunk! Tha-thunk! How long are these things going to last?!
It was over 400 miles to the New Mexico state line.
Tha-thunk! Tha-thunk! Could they stretch that far?!?!
I tried to stay calm and focus on pedaling, but the gouges were relentless.
Tha-thunk! Tha-thunk! Tha-thunk! Tha-thunk!

Five miles of violent jarring broke me. I jammed on the brakes, dismounted, and shoved the bicycle to the blacktop before punting my helmet onto a rocky hillside. Marching furiously along the shoulder with my right arm and thumb extended, I wanted to hitch to the nearest bar and begin my final undoing.

Fortunately, there wasn't jack shit—cars or buildings—for twenty miles in any direction. I slumped on the guardrail to regroup. Surprised to have a cell signal, I searched the web for information about the gouges but found nothing. I considered riding on the grass that abutted the breakdown lane but realized that wasn't a 400-mile solution. As a final hope, I dashed across speeding traffic and inspected the eastbound breakdown lane, my plan to ride it west if it was gouge-free. Sadly, it was equally marred.

I recrossed the interstate and shuffled back to my bicycle. If I knew how far the gouges stretched, then I could plan accordingly. The prospects of continuing without that crucial information were bleak, and I thought about quitting. If not that, then at least hitching some distance to minimize my misery. Sonia was always a calming influence, so I called her for advice. When she didn't pick up, my spirits sank further.

A minute later, while I was weighing awful options, my phone alerted me to an incoming email. It was from Tom, a birder in Indiana. He explained that he stumbled into my blog two weeks ago and had since read all the earlier entries. His note concluded, "Reading about your travels is a wonderful distraction. I can't wait to see how the rest of your adventure unfolds. Best of luck as you go for 600. I'll be right there with you!"

At that instant, any thought of quitting or hitching evaporated. Readers like Tom had been supportive since my first mile, and I knew I had to keep the adventure rolling. I retrieved my helmet and climbed into the saddle, my blog audience lending motivation as I renewed my misery.

My joints throbbed, my teeth chattered, and my backside burned as I bounced along the westbound shoulder. I rode in the right-hand lane when traffic permitted, but that was only possible for short stretches and discouraged by the deafening honks from eighteen-wheelers as they overtook me at eighty miles per hour. The distance between each milepost felt like a hundred, and I cursed each impact as I battled the unyielding landscape.

Tha-thunk! Tha-thunk! Tha-thunk! Tha-thunk!

After ten torturous miles, the gouges vanished. Optimism renewed—*Arizona, here I come!*—I accelerated to my usual cruising speed of fourteen miles per hour. Stretches of gouges surfaced through the afternoon, each inducing a temporary panic attack, but none lasted for more than a few hundred yards. Had they, I might have steered into oncoming traffic and been done with it.

I reached Sonora, Texas, that evening and gained Ozona, Iraan, Fort Stockton, Balmorhea, and Van Horn in subsequent days. With midday temperatures topping ninety degrees and headwinds building into each afternoon, I began my rides two hours before sunrise to minimize those complications. West Texas is free of light pollution, and the stars shone like a million twinkling eyes as I turned my pedals under their nocturnal gaze. Passing cars illuminated the desert beyond the meager reach of my bicycle light, and I tried to imagine what early-morning drivers thought as they zoomed by me on one of the most desolate stretches of road in the country.

Sunrise warmed my back and painted the rocky landscape with soft pastel hues. Juniper and mesquite yielded to scrubbier vegetation as I reached the higher elevations of the Chihuahuan Desert, and earth tones merged with azure sky in a shimmering midday haze that obscured the distance ahead. A lone rider in the endless Lone Star state, I found strength knowing I'd outlasted bumpier times.

I began graduate school at New York University in August of 2004 and had zero difficulty balancing the coursework and laboratory rotations characterizing the first year of a biomedical PhD. While my classmates studied, I partied. Within a month, I'd established a routine featuring blackout drinking three or four nights each week. That I attended classes reeking of booze and cigarettes didn't seem to matter; I grasped concepts and discussed papers as well as any of the sober students.

Outside NYU, my music interest introduced me to DJs, promoters, bouncers, and clubland socialites; those contacts lifted velvet ropes, minimized the monetary cost of partying, and helped me secure DJ gigs at several bars and lounges. I also ran a series of illegal loft parties at a photography studio in a commercial building near Union Square. I DJed and pocketed the $5 cover, and the photographer used his well-stocked speakeasy to defray his exorbitant rent. Several hundred people attended each monthly event, so it was consistent money for both of us.

Flush with cash—and my tuition paid by government grants—my cocaine use exploded. Dealers were willing to meet me anywhere at any time, so scoring blow in New York was easier than ordering pizza. On Saturdays and Sundays, the drug kept me bouncing between bars, clubs, and loft parties until midday, and I began bowing to conditioned cravings in the course of my midweek drinking. Tuesday happy hours became all-night benders, and I routinely attended class under the residual influence of cocaine, usually on little to no sleep.

Despite the fact that graduate school was not my priority, I survived my first year in strong standing while taking an interest in cell polarity, a fundamental process through which cells regionalize function. A neuron is an example of a polarized cell; it receives electrochemical input on its dendrites and outputs the signal using its axon. Likewise, an intestinal cell absorbs nutrients on its apical side and dumps them into the blood-stream from its basal other. Many cellular functions are directional, and understanding how that vectorial asymmetry, that polarity, is regulated has major implications for development and disease.

A plan for my doctoral dissertation emerged; I'd employ *Caenorhab-ditis elegans*, a microscopic worm that has facilitated countless biological breakthroughs, to identify novel polarity regulators. Polarity genes were known to function across the evolutionary spectrum, so any additional molecules I identified in the genetic model would inform human biology as well. The question framed and the system defined, I'd need five years to engineer the animal's DNA, perform experiments, collect data, and publish findings.

That process would unfold under the guidance of my advisor, Dr. Jeremy Nance. Aged thirty-six, Jeremy was young by faculty standards. He was a recent hire after his postdoctoral fellowship at the Fred Hutchinson Cancer Research Center in Seattle. I was only the second person to join his nascent polarity group—the other a research assistant—so Jeremy trained me in the absence of intermediaries. That we worked so closely facilitated the discovery of an unlikely connection.

"I can't shake the feeling we've met before. Are you a birder?" Jeremy asked.

"Used to be," I said.

He continued, "Have you ever birded on the Outer Banks?"

"Yeah. My friend George Armistead dragged me down there in 1998 to help with a winter census," I replied. "I'd mostly lost interest by then—my second year at Stanford—but maybe we met at the dinner which followed."

We didn't pin down the particulars of that coincidental intersection, but we grew our birding connection in ensuing months. We coordinated outings to Central Park and Jamaica Bay when I could abstain from drinking, and we initiated a joint, five-boroughs bird list that relied entirely on public transportation. The exercise breathed life into my comatose interest and bridged the advisor–student gap in an unusual way.

Unfortunately, my swelling substance abuse had the opposite effect on our relationship. Jeremy was aware that I routinely arrived at his lab with bloodshot eyes and alcohol on my breath, but he waited until I missed an important meeting with another lab to confront me.

"What happened to you this morning?" he asked.

I tempered my answer to omit the details, the meeting long forgotten by the time I blew my last line of cocaine at five a.m. "Damn, I completely forgot," I said. "Got a little carried away last night. Slept right through my alarm. I'm sorry. It won't happen again."

That promise was as empty as the dozen beer bottles on my coffee table, and I missed another meeting a month later.

Jeremy was exasperated. "You've gotta get control," he said. "I didn't say anything when it was just you beating the hell out of yourself, but it's beginning to affect our relationships with other labs."

I replied sheepishly, "I know. I've let myself go. I'll get a handle on it."

"You need to do more than that," he said assertively. "You have a drinking problem, and you need to talk to someone about it."

My confidence departed me in that instant, the weight of his words seemingly lowering the adjustable-height chair on which I was sitting. Friends and acquaintances had hinted at that suggestion for years, but no one had voiced it so explicitly. Simultaneously exposed on personal and professional grounds, I felt like a turtle without a shell. Rather than get defensive, I told Jeremy what he wanted to hear.

"I'm going to put drinking on pause and think about the role that alcohol is playing in my life," I said. "I know I have to make better decisions moving forward."

Jeremy didn't impose an ultimatum, and I limited my drinking in subsequent weeks, though I did that more to avoid potential friction than to initiate a personal inventory. I performed a very informative experiment a month later, and a subsequent avalanche of exciting data lent the cover that allowed my behavior to return to its elevated baseline. Periodic screwups sowed temporary conflict and regret, but Jeremy was handcuffed because I always delivered on the experimental front. Just as at previous scholastic stages, I thought my academic performance forgave, even justified, my self-destructive behavior. I simply hadn't paid a sufficient price to motivate behavioral change—yet.

The cycles I spun through with Jeremy were reminiscent of those I experienced with my family. I'd get hammered with my neighborhood cronies when I visited my parents in Philadelphia, and they labeled me "out of control" whenever I stumbled in at five a.m. or passed out on the floor. Fortunately, their glimpses were limited and minimized against continued academic progress.

Given their previous views of my drinking, I was shocked when my parents called me to discuss my cousin's substance abuse issues. It was November of 2006, and I was a few months into my third graduate year. Alexis and I got along—and got drunk and high—on the rare occasions when we got together, but I wasn't aware she'd established behavioral patterns more self-destructive than my own because we rarely communicated, each of us prioritizing our relationship with booze over familial connection.

Frustrated with her dodging, Maureen and Bob, my mom's sister and brother-in-law, had hired a private detective to trail Alexis for two weeks. His resulting report indicated the situation was dire; she lived in a condemned flophouse, associated with known lowlifes, and often drove erratically, presumably while drunk and high. A professional counselor had deemed a formal intervention necessary, and the family wanted me to be a part of it. It was an odd request given what they knew about my drinking—and an insane ask given what they didn't know about my drugging—but I agreed to meet them in Pittsburg at the beginning of December.

I arrived at the hotel suite to find Aunt Maureen and Uncle Bob in the seating area, along with my parents and younger sister. We'd each prepared a letter voicing our love and concern for Alexis. Kurt, the professional counselor, offered a brief outline of how the intervention would unfold. Our only goal was to persuade Alexis to accept the help we were offering; if she refused, then we'd sever communication until she agreed to get clean and sober.

Lured to the hotel under the premise of brunch with her parents as they drove from Philadelphia to Cleveland, Alexis knocked on the door fifteen minutes later. She entered glassy-eyed and immediately realized what was happening. "Ahhh, yes. Let the healing begin," she slurred as she sat between her parents.

Bob's letter didn't dent her alcoholic armor, and my parents' offerings hardly registered. Alexis deflected and denied throughout, so the family counselor invited me to read my prepared note. My outreach precipitated an expectedly strong defense.

"Are you fucking kidding me?" she exclaimed. "You get more fucked up than I do!"

I countered, "I know, but I'm in school and making progress toward something. You're unemployed, drifting, and in danger of ruining your life."

That judgmental rebuttal incensed her. "Career plans mean you're not an alcoholic? Got it," she shouted. "For a smart guy, you say some really dumb shit."

I replied, "Today is about you, not me."

She exploded. "You want a fucking medal? This is just some bullshit so you can feel better about yourself."

"Listen, Alex. You're in some deep shit, and we don't want it to get worse," I said.

Kurt stepped in and tried to regain control of the intervention. "Everyone here loves you, so please consider what they're saying."

Alexis erupted in laughter before sashaying out of the room. I ran after Alexis and caught her in front of the hotel.

"Don't do this, Alex," I begged. "I've got a few cigarettes left. Give me that long."

Bowing to chemical dependency, she acquiesced. We found a bench near the hotel entrance. I lit a cigarette and handed her one. She lit it, took a deep drag, and exhaled. "So, you're really going to do this?

"Look, our parents have given us both a lot of leeway. But you showed up wasted to what you thought was brunch. Not dinner. Not a baseball game. Not a cocktail party. Brunch. If you'd shown up sober, you might have a leg to stand on. But this was the one time you couldn't get fucked up, and what did you do?"

"I got fucked up."

It was the first responsibility she'd claimed. The tension lessoned through the next few minutes. She voiced apprehension: "I'm not sure I can do it."

"Just try and see where you end up. Alcohol and drugs aren't going outta style. They'll be here if you decide sobriety sucks. Trying and failing is better than not trying at all."

"OK," she said quietly.

"OK, what?" I asked.

"OK, I'll try it," she replied.

Avoiding her hair with my cigarette, I leaned over and hugged my cousin. I knew she could change her mind at any moment, so I handed her another cigarette as a distraction and ran inside to fetch the family.

They rushed outside, and Maureen and Bob loaded Alexis into their car and whisked her toward treatment. I had a quick lunch with my family, references to my drinking thankfully avoided, and caught an afternoon train back to New York. I tried not to think about the day's events on my return trip; doing so would have forced me to face Alexis's warranted accusations.

I got blackout drunk later that night. And again the next. Waking in my own piss on that second of consecutive mornings, my hypocrisy became self-evident. When I should have walked into sobriety alongside Alexis, I'd forced her in that direction while standing still. Mine was an offensive display in front of my closest family, and I couldn't think of any way to assuage the guilt I felt besides getting sober. That night, unbeknownst to everyone, including Alexis, I walked into a church basement for my first Alcoholics Anonymous meeting. I was finally in the right place—for all the wrong reasons.

My journey into sobriety would prove as long and bumpy as my Texas traverse, that two-week ordeal ending when I crossed into New Mexico north of El Paso midday on May 12th. Since departing the Gulf Coast, I'd cycled 900 miles through some of the most desolate and inhospitable terrain on the continent. I'd outlasted oppressive heat, survived swirling gales, and tackled my first extreme topography. The challenges and logistics were herculean, but I overcame everything Texas threw at me, gouges included. Arizona's birding riches within reach, I turned my attention to southern New Mexico.

FOURTEEN

Past, Present, and Future

Wheeling across New Mexico after completing my Texas traverse, I overnighted in Las Cruces on May 12th before reaching Deming and Lordsburg in subsequent days. I departed that last destination on May 15th and continued west on Interstate 10 before abandoning the transcontinental at Road Forks, a collection of mostly abandoned buildings six miles short of Arizona. Veering south onto Highway 80, I navigated tire-pinching cracks and rim-bending potholes as I paralleled the state line through the desert. Heat waves collided like ocean breakers, animating the parched landscape in dizzying and disorienting ways, and I experienced something akin to seasickness as I powered through the parched landscape

A flat tire brought me to a grinding halt, the oppressive heat assaulting me in the absence of the airflow peddling provided. Sweat poured down my salty brow and stung my eyes, and lightheadedness rendered me clumsy as I fumbled with tools and parts. I reached for my water bottle, but its lukewarm contents didn't refresh. The tire repaired and my transport reassembled, I resumed forward progress, self-generated airflow making the unbearable bearable once again.

I departed Highway 80 outside Rodeo, New Mexico, and steered west into Arizona, toward the Chiricahua Mountains. Rising steeply out of the desert thirty miles north of the international border, the range hosts an avifauna that suggests Mexico as much as the Mountain

West. Upward of twenty Latin American species reach their northern limits in the mountains of southeastern Arizona, and the Chiricahuas are a popular destination for birders seeking those mostly foreign birds within the United States. My plan was to spend five days in the range and another ten scouring additional birding hotspots along a roundabout path to Tucson before turning north, toward Flagstaff and—eventually—the Rockies.

The midday heat clung to me like a freeloading passenger as I struggled off the desert floor and gained elevation toward Portal, a tiny town at the mouth of birding-renowned Cave Creek Canyon. I paused for lunch at the Portal general store, and I gave my feverish hitchhiker the heave-ho in the rejuvenating shade of oaks and sycamores in the higher throws of the massive drainage.

There was an avalanche of new birds in those cooler surrounds. A raucous flock of blue-gray Mexican Jays foraged at the roadside; an alto Black-throated Gray Warbler sang sweetly from the top of a nearby pine; an acrobatic Brown-crested Flycatcher plucked an insect from midair. After two weeks of grueling riding, Arizona's birding riches were the best possible reward.

Among many wonderful birds, none matched Acorn Woodpeckers for entertainment value. Inhabiting arid woodlands across the southwestern United States, the black-and-white harlequins are—as their name suggests—fond of oak nuts. Each family of two to fifteen birds caches thousands in custom-drilled holes in the sides of trees, and there is constant grunting and bickering as neighboring clans vie for ellipsoid edibles with which to fill their respective pantries. Watching the rivals squabble, I recalled how as kids, in the days following Halloween, my sister and I would sneak into the other's room to steal candy. Birds often exhibit behavior that reminds me of my own humanity, and I smiled as the woodpeckers battled over nature's equivalent of a Reese's Peanut Butter Cup. At that moment, the immense evolutionary distance between the woodpeckers and me felt trivial.

Continuing up canyon, I reached the bucolic Southwestern Research Station at 5,400 feet of elevation. I dumped my gear in my assigned cabin, collapsed onto a plastic lawn chair on the porch, and oriented myself toward a nearby hummingbird feeder. The action initiated a moment later when a winged dart whizzed past my face and landed on the red faux-flower feeder. It was a male Blue-throated Hummingbird; at five-and-a-half inches long, he was a giant, his cobalt gorget glistening as he lapped sugar water from the dangling dispensary.

A purple-green bolt rocketed toward the same buffet. It was a male Magnificent Hummingbird, another behemoth. Irked by the presence of his Blue-throated rival, he'd arrived to defend his favored food source against intrusion. The ensuing clash was as captivating as it was heated, the iridescent aerialists twisting and contorting as they fought to establish midair advantage. The pair flew three circles around each other in the next instant; I couldn't discern which wings belonged to which combatant because they were so closely engaged. The giants occupied, a smaller Broad-tailed Hummingbird cautiously approached the feeder. The opportunist sipped for a few seconds before being bounced by the Magnificent, the newly minted heavyweight champion.

Hummingbirds are one of evolution's most incredible creations. More dynamic than dainty, they are some of the most accomplished aerialists in the avian world and uniquely capable of backward flight. An average-sized, four-inch hummingbird flaps its wings eighty times per second; its heart—smaller than a kernel of corn—beats over a thousand times per minute. Their biology is trumped only by their beauty and assertive personalities, and I could have spent all afternoon watching the hummers if other birds didn't demand my Big Year attention.

Exploring around the research station, I added Violet-green Swallow, Cassin's Kingbird, and Yellow-eyed Junco to reach 400 species, that benchmark a satisfying measure of the physical and mental challenges I'd overcome to achieve it. I'd covered 6,070 miles of an anticipated 15,000—the equivalent to mile 11 in a marathon of 26.2—and I was optimistic I could maintain my motivation and health as the year progressed. Even if circumstances stopped me short of my 600 species goal, I knew I'd already found the most important thing: the will to depart a scientific course that would never have satisfied me. Compared to the constricted view from my laboratory, my wider Chiricahua perspective was wonderful.

That night I took an extended stroll along the dirt road that delivered me to the research station. The air was crisp, and I could hear the gentle toots of Elf Owls from nearby trees and the haunting whistles of Mexican Whip-poor-wills from the overhead slopes. I turned off my flashlight and sat on a roadside log, my heart slowing and my aches abating as I took a series of deep breaths.

I recalled my first visit to Arizona, twenty-three years prior. Age twelve and attending Camp Chiricahua, a two-week immersion for bird-obsessed kids, I was too young to think beyond the birds I was observing. I knew nothing of evolution or conservation, and I couldn't have imagined that I'd eventually pursue my doctorate in molecular genetics. My first drink was still five years away, and the idea of alcoholism was as incomprehensible as the notion of romantic love. Too inexperienced to imagine my life's arc, I lived each day free of ego and expectation, my future more a blank page than a to-do list.

Despite that limited worldview, I knew that birding had irrevocably affected me. The pursuit was an addictive blend of observing, collecting, and learning, and I saw the endeavor as a never-ending treasure hunt, one that would occupy me well into old age. I didn't know if birding would remain a mere diversionary pursuit, suggest a course of academic study, or grow into a vocation, but I knew the passion would be a part of me as long as I was above ground.

Seated on the Chiricahua log, alone in the Arizona darkness, I realized birding had assumed a significance I never imagined; it was my most-enduring personal thread, my strongest connection to the wide-eyed child who'd walked these same roads and trails two decades ago. Overshadowed by academics and overrun by alcohol, the interest had survived. Now a resuscitated passion, it prescribed my transcontinental trajectory amidst my thirty-something confusion. I'd been too preoccupied with unwinding the past and managing the present to think about the future since I started pedaling, but I wondered what role birds and birding would play in my life beyond the conclusion of my adventure. I realized that the best way to maintain my passion would be to share it with others, perhaps as a birding guide or a conservation advocate, but I still had to survive another 230 days on the bicycle before any of the particulars could take shape. Birds and birding made me happy, and at that midnight moment, that was enough. Reassured that the future would resolve in due course, I retired to the research station and passed out.

Amidst fantastic Chiricahua birding the following day, May 16th, I prepared to rendezvous with Ron Beck, an Arizona birder with whom I'd been communicating for several months. Ron discovered my blog in January and immediately pledged his support when—and if—I reached Arizona. Retired and living with his wife, Janet, in nearby Hereford, Ron was then only weeks removed from his own bicycle Big Year, a 2013 effort during which he cycled 3,000 miles and found 301 species in his home county, Cochise. That he accomplished that feat at the age of 59 was impressive; when most his age were slowing down, Ron was plowing straight ahead, an observation further supported by his fondness for extreme sports and snake handling.

Ron and Janet met me at the research station on the morning of May 17th. His muscular frame, sharp eyes, graying goatee, and silver hoop earring contrasted with her petite, blonde, doe-eyed presence. They immediately produced three sturdy mountain bikes from the bed of their pickup truck.

"What's up with the extra bike? Someone else coming?" I asked.

Ron replied, "Nope. It's for you. Yours doesn't stand a chance on the rough road."

His offer was gracious, but I was apprehensive about using the loaner. "I dunno. I'm kinda attached to mine after five months. I feel like we're in it together."

He laughed in my face and replied, "You biked what, like 5,000 miles to get here?"

Setting the record straight, I replied, "Just over 6,000."

He'd set the trap perfectly. "Well then, twenty miles on the loaner doesn't amount to shit, now does it?" he asked rhetorically.

His point made, I mounted the loaner, and the three of us started the 3,000-foot climb toward Rustler Park at 8,500 feet, where we planned to search for several high-elevation species. The mountain bike was a good decision; without its suspension system, the rutted and rocky road would have replayed my gouge misery ten-fold. Any apprehensions I might have had about Janet were also dispelled; like a hummingbird, she was a powerful cyclist in a tiny body.

Cranking along in our lowest gears, we made slow but steady progress up the Forest Service track. Deciduous forest yielded to conifers, and we found Greater Pewee, Pygmy Nuthatch, Steller's Jay, and Olive Warbler as we ascended. Most exciting were two Mexican Chickadees; common in the mountains of Mexico, the species is restricted to the highest elevations of the Chiricahuas within the United States. The bird was the prime motivation for our climb and capped a productive morning.

We descended after lunch. Ron and Janet rocketed down the mountainside with an abandon I couldn't believe. Where they steered a fluid course, I navigated individual bumps and rocks, my nervous wobbles delivering me to the research station twenty minutes behind them. They were sacked out on the grass when I arrived, but I roused them to load the bikes into their truck. As we did, I mentioned that my forearms were tired from riding the brakes down the mountainside. Ron looked at me and laughed.

"What?" I asked curiously.

"Brakes are for chumps!" he bellowed.

It was trademark Ron. Despite his gruff, tough-guy exterior, he was as sincere as anyone I'd met in my travels. His knowledge of Arizona natural history was extensive, and his ritual commitment to health and fitness was inspiring. I envied the simplicity of retired life and the loving relationship he'd cultivated with Janet. For his part, Ron gushed over my endless energy and enthusiasm for biking and birding. In Ron, I saw a preview of my future; in me, he saw a reminder of his past. Our interaction exemplified the ageless connections birding facilitates, and reciprocal man-crushes guaranteed we'd be friends for life. He and Janet departed after goodbyes, and I did a bit of late-afternoon birding ahead of dinner, blogging, and bed.

I spent two additional days kicking around Cave Creek and Portal before vacating the Chiricahuas on the morning of May 20th. None among Montezuma Quail, Common Black Hawk, Elegant Trogon, Whiskered Screech-Owl, Buff-breasted Flycatcher, and Red-faced Warbler escaped my blazing binoculars, and my five days in the

Chiricahuas yielded 49 new species to push my total to 431. My visit to the remote range was a welcome reminder of my past, and I knew I'd miss the nostalgia as I dropped into the searing desert and rejoined Highway 80 south toward Douglas.

In subsequent days, I visited most of the traditional birding hotspots in southeastern Arizona: the Huachuca Mountains, Patagonia, Sonoita Creek, Madera Canyon, and Mount Lemmon. It was blisteringly hot at many points, but productive birding redeemed the discomfort, with Rufous-capped Warbler, Black-capped Gnatcatcher, Buff-collared Nightjar, and White-eared Hummingbird making celebrated appearances. By sunset on May 30th, I claimed 457 species, a five-month total exceeding any prediction I'd made from Massachusetts. Even better, I was running ahead of schedule; my premature departure from the Gulf Coast had coupled with an expedient Texas traverse to create a valuable week of cushion. Unsure how my body would hold up across the remaining seven months of the year, I was encouraged that I could afford myself rest if fatigue, sickness, or injury befell me.

I'd already explored a lot of the country, but my exhausting migratory riding between New England, Florida, the Texas coast, and southeastern Arizona was offset by stationary time in each of those destinations. However, my approach would change once I reached the Rockies because there wasn't one location—or even a handful of locations—where I could find large numbers of the mountain birds I hoped to see. I'd need to pursue species individually, making the Rockies not so much a singular destination as a continuous ten-week grind. My high-elevation summer would present the most demanding riding I'd encountered, and I'd be lucky to add another 60 species before I intersected the Pacific Coast in Oregon, hopefully in early September. Departing Tucson on the morning of May 31st, I steered north toward canyonlands. I was ready to face the mountains. All I had to do was survive the dogs.

FIFTEEN

Fire and Ice

The attack commenced without warning, when two seething German Shepherds exploded from the brush as I wheeled north out of Tucson, Arizona, on May 31st. I'd outlasted dozens of canine pursuits across the previous five months, dogs of every description tearing into the street in pursuit of my bicycle as I explored rural areas, but the latest pair came at me with unparalleled fury. They were barking and snarling with each powerful stride, and I knew I was fucked when their menacing eyes saw the gathering fear in mine.

My fight-or-flight response ignited, and my heart started thumping like a jackhammer inside my chest. I tried to outrun the growling assailants, but the convoluted bicycle path prevented a straight-line escape, the attackers gaining ground as I slowed around each bend. Unable to outstrip them, I waved my right arm above my head while steering with my left. That first-line defense had proven sufficient to keep most dogs away from my moving bicycle, collisions more of a threat than bites, but the bloodthirsty beasts intensified their onslaught in response to my gesture. When second-line squirts from my water bottle didn't stymie their blitz, I unclipped my feet from my pedals and prepared for battle.

I struggled to extend my faltering escape against a coordinated effort to bring me to a stop. Accelerating ahead of me, the sacrificial smaller dog positioned himself directly in my path while the trailing larger lunged at my right leg with fully brandished teeth. I skillfully avoided the former while swinging my lower appendage away from the latter, but the velocity I surrendered left me exposed to a second round of block-and-bite a moment later. With my legs deployed in defense instead of pedaling, the attackers brought me to a near-standstill in thirty seconds.

I didn't think I stood a chance if I was forced from the bicycle, so I used all my strength to remain upright and rolling. That task became harder each instant, hyperventilation rendering oxygen in suffocatingly short supply. I scanned the edge of the bike path for a stick or bottle I could use to defend myself in the event I was grounded. I saw a grapefruit-sized rock fifty feet ahead and set my sights on it, my Big Year—and possibly my life—hanging in the balance.

The trailing dog lunged at my right leg again, and I met his assault with a swift kick. The blow struck his face with a resonant thump but didn't dent his resolve. A forceful boot to his neck on his next lunge yielded the same, ineffective result, and it seemed I was facing an indestructible opponent, the canine equivalency of The Incredible Hulk, Rocky Balboa, and Keith Richards rolled into one. My leg was cocked when another strike came, and I landed an uppercut. He yelped, recoiled, and yielded, and I used a left-footed body-blow to back the accomplice down. My escape cleared, I gathered momentum, bypassed the rock, and never looked back.

Despite dozens of previous encounters with dogs, none had escalated to attack level. I survived without injury but had likely hurt the larger dog, a realization precipitating a mix of relief and guilt as my adrenaline equilibrated. I hated being aggressive toward any animal—dogs usually the most lovable creatures on Earth—but I had no choice under the circumstances. I knew self-preservation was an acceptable reaction, but the episode weighed heavily on me through the remainder of my ninety-mile ride to Casa Grande.

I reached the Phoenix inferno the next day. The 106-degree hell-fire was miserable even when I was moving, and endless traffic lights ensured that I had plenty of stationary time to broil. The heat wrung sweat from my body as though I were an old washrag; no amount of Gatorade could stave off dehydration and resulting leg cramps. By day's end, my shirt had absorbed so much excreted salt that it crinkled like Bubble Wrap when I removed it from my increasingly emaciated body.

I powered north to Flagstaff via Black Canyon City and Camp Verde in the next three days. Covered mostly along the shoulder of Interstate 17 (not to be confused with Highway 17 through the Carolinas and Georgia), those 150 miles featured the most challenging topography that I'd faced and netted 6,000 feet of elevation gain from 10,800 feet of climbing—the equivalent of seven-and-a-half Empire State Buildings. Cooler temperatures offered reprieve as I departed desert scrub, ascended into pinyon-juniper, and explored Ponderosa pine forest, but the unyielding uphill introduced me to new brands of fatigue.

Departing Flagstaff after a two-day recovery pause during which I added Lewis's and American Three-toed Woodpeckers, my total then 464 species, I followed Highway 89 as it dropped 3,000 feet onto the western edge of the Painted Desert. Even under harsh midday light, the landscape was gorgeous. Every conceivable shade of red, orange, and pink was represented in the stratified geology, and the ever-shifting hues imbued the rocky roadside with magical qualities. Mirages made the blacktop more a mirror than a highway, and I wondered if what it reflected was more revealing of the past or suggestive of the future. Maybe my mind had been invaded by desert spirits; more likely it was dehydration-induced hallucination. The intrigue was palpable, but the scientist in me refrained from rationalizing it one way or the other. The feeling was more valuable than an understanding of it. Deliriously lost between land and sky, I pedaled deeper into Navajo Nation.

After overnights in Tuba City and Kayenta, I followed Highway 163 north into a sweeping valley harboring a majestic array of rock formations. Behemoth and brick red, the gargantuan mesas rose hundreds of feet above my head as I navigated their exaggerated, early-morning shadows. I'd seen pictures of Monument Valley but never imagined its grandiose scale, my progress insignificant in the context of its overpowering geology.

I tried to fathom the eons required to sculpt such prehistoric formations as I explored the negative space between them. Contrary to what intuition might suggest, they were neither purposefully raised from the underlying earth nor divinely dropped from the overlying heavens, erosion instead carving the relics from a uniform plane over hundreds of millions of years. Time and pressure are the ultimate sculptors, and I crawled into Utah in awe of their Monument Valley masterpiece.

My brush with the Beehive State was brief, and I crossed into Colorado at Cortez before reaching Durango on June 11th. There I took a day to rest ahead of my first confrontation with the Rockies, a monstrous, 5,300-foot climb to reach Silverton. Following Highway 550 after my recovery pause, I paralleled the Animas River through the lowest reaches of the San Juan National Forest. Family farms abutted the meandering flow for the first few miles, but conifer-covered slopes prevailed as I climbed, the cheerful chirps of Mountain Chickadees filling the air as I cranked toward Coal Bank Pass at 10,640 feet.

The first nineteen miles of that ascent assumed a manageable grade, but the pitch steepened across the final six. Available oxygen decreased as I reached above 9,000 feet, and my pace slowed to a crawl under anaerobic conditions. Lactic acid stung my legs with increased intensity at each milepost, and I turned to trees as a finer measure of my sluggish advance. Coupled with intensifying overhead sun as the morning progressed, the summit seemed impossible.

After three hours of uninterrupted grinding, the road leveled off, a wooded picnic area came into view, and a sign announced my arrival at Coal Bank Pass. I was exhausted and limped into the adjacent pullout to rest and rehydrate. Though 10,640 feet was the highest elevation I'd reached by more than 2,000 feet, my satisfaction was momentary; with the higher Molas Pass (10,910 feet) looming, there was more pain ahead.

Continuing down the backside of Coal Bank Pass at speed, I surrendered a thousand feet of elevation in the next three miles. My gravity-aided euphoria was short-lived, and I was soon struggling up another steep, five-mile pitch toward Molas Pass. Already pulverized, my legs felt like marshmallows as I pushed skyward. My breathing broke down, and discontinuous heaves and shallow flutters replaced rhythmic cycles. Inches felt like miles, but I ignored the burning desire to stop and rest. As long as I was moving, I'd stay ahead of the gravitational beast trying to wrap his scaly fingers around my rear wheel.

The trees slowly thinned, and a sweeping alpine panorama unfolded as I rounded the final bend and crested the pass. Lush meadows abutted the road and cascaded downslope in undulating leaps and bounds; dotted with wildflowers and lichen-covered boulders, they merged with a thick coniferous forest below my vantage. Molas Lake reflected the azure sky, and an intimidating collection of snow-capped peaks, the stoic gatekeepers to the mountainous world beyond them, dominated the distance. The day's climbing accomplished, I lolled at the overlook while soaking in the scenery.

I couldn't believe I'd departed the desert just five days prior. Had I not witnessed the transition from my bicycle, I could have believed the dichotomous scenes were from two distant planets, one forever burning and the other forever frozen. I'd experienced the American West previously, but I never felt so much a part of the landscape as I did on the bicycle. Without air conditioning, my sweat dripped onto Monument Valley's scorched earth; without an engine, my legs delivered me to the mountaintop. All previous views of the world suddenly seemed so sterile, so safe, and a more authentic view of the landscape emerged, one suggesting the West was still as wild and untamed as it had been centuries earlier. My spirit invigorated, I began the seven-mile, 1,600-foot descent into Silverton, that veritable free-fall taking barely fifteen minutes. On one straight stretch, I leaned forward and lifted both legs off the pedals; it was the closest I'd ever felt to flying.

As the name suggests, Silverton was founded as a mining outpost in the late nineteenth century. Lodes of gold and silver sustained the town for well over a hundred years, but tourism has replaced precious metals as the town's economic lifeline. The surrounding mountains have kept the quaint municipality trapped in time, and gravel boulevards and old-fashioned storefronts recall the frontier era. The fully operational Durango–Silverton railroad was particularly exciting, a nostalgic throwback to an era when steam and the pioneer spirit—rather than the microprocessor and corporate influence—dictated the nation's future.

Strolling Silverton's streets, I fully expected some white-bearded fossil loafing in front of a local saloon to point to the surrounding mountains and ask if I'd "found any gold in them thar hills." At other moments, preppy couples pushing designer strollers suggested the town was a typical tourist trap. The blend of past and present worked, and I was disappointed to leave Silverton as I began the heavy climb out of town two days later, the intervening day used to add two high-elevation finches, Red Crossbill and Pine Grosbeak, to my growing list.

Continuing on Highway 550—appropriately dubbed the Million Dollar Highway—between Silverton and Ouray, I summited Red Mountain Pass at 11,017 feet before dropping onto the most spectacular 13-mile stretch of road I'd experienced. The top section of the 3,300-foot descent was as loopy as cooked spaghetti, and I leaned aggressively into each hairpin to maximize the centripetal force and the associated adrenaline rush. With an alpine world rotating around me from each sweep of my handlebars, I felt like the center of the universe as I slalomed down the mountainside.

The road straightened. I crouched to gather speed. Roadside aspens blurred as I rocketed down the drainage, and my heartbeat accelerated in proportion to the rush of alpine air across my ears. Exceeding forty miles per hour without any effort, I'd found my happy place.

A second twisted stretch snapped me out of my temporary trance, and I reestablished focus while navigating overhanging rock on one side and a thousand-foot plunge on the other. There was no room for error, and I held my breath as I whizzed past ascending cars and around blind turns. It was a death-defying course, a roller coaster ungoverned by protective harnesses or emergency stop buttons.

I departed those craggy confines and emerged onto the lateral slope of another sweeping valley, a small town visible at the bottom. Continuing my plunge, I snaked through a final series of twists and turns before rolling into Ouray. A mining-turned-tourism municipality in the Silverton image, the outpost afforded access to my next Rocky Mountain target, the Black Swift.

Superficially similar to swallows, swifts are most closely related to hummingbirds; with bodies like carrots and wings like sickles, their aerodynamic profiles make defense contractors drool. Though birds of prey like the Peregrine Falcon reach greater speeds in gravity-aided dives, swifts are the fastest straight-line fliers in the world, some exceeding

100 miles per hour. They possess incredible endurance; telemetry studies have revealed that some species stay aloft for months at a stretch. Swifts feed on airborne insects and drink by skimming across rivers and lakes, and extraordinary examples have been observed copulating in midair, they the founders of the vaunted mile-high club.

At seven inches long, the Black Swift is the largest of four North American swifts, the elusive bird nesting in isolated Rocky Mountain and Pacific Coast pockets before migrating to the tropics for the winter. I could theoretically intersect the indefatigable bird anywhere in the Western United States, but I'd set my sights on Ouray because it hosts the most accessible breeding colony on the continent. Faced with the choice of riding extra miles to pursue a particular species at a reliable location or hoping for a chance intersection along a shorter course, I always chose the former because riding miles were more calculable than bird behavior.

I arrived at Box Canyon Falls early in the evening. I ditched my bike and walked across a metal catwalk into a huge fissure in the cliff face. Cascading snowmelt produced a thunderous roar as it escaped an overhead crevice and crashed onto the underlying rocks, and I refreshed myself in the cool mist that filled the immense cavern. Against the overpowering waterworks, most visitors don't notice the swifts' rudimentary nests; made from moss and saliva, the donut-sized platforms clung to small ledges on the spray-soaked canyon walls. No eggs or young in evidence by my June 15th arrival, I'd need to wait until the foraging adults returned at sunset.

I exited the fissure and assumed a position that afforded me views of the sky. The sun sank, and a dozen swifts appeared high overhead as the heavens turned purple. The feathery boomerangs appeared reluctant to resign their wings, their nimble forms knifing through the twilight with a blend of power and elegance, but approaching darkness forced their aerial acrobatics lower. Looping and cutting just meters from my disbelieving eyes, they made final passes of the cliff face before winging into the misty fissure. My subsequent views of the birds on their nests were intimate but anticlimactic; once I'd seen them flying, a static snapshot would never compare. Night consumed the canyon, and I retired to my motel. Like the swifts, the rising sun would render me kinetic once again.

Over the course of the next ten days, I wound north and east through Montrose, Gunnison, Salida, Buena Vista, and Silverthorn. Among Gray Flycatcher, Willow Flycatcher, Clark's Nutcracker, and MacGillivray's Warbler, the endangered Gunnison Sage-Grouse was my best addition along that route. Its range restricted to southeastern Utah and southwestern Colorado, it was a must-find before I continued northeast.

By the time I reached the Front Range west of Denver, I claimed 476 species. There were, unfortunately, still two glaring holes in my high-elevation resume: White-tailed Ptarmigan, a small grouse, and Brown-capped Rosy-Finch, a chunky songbird. Both species spend the summer above 12,000 feet, the ptarmigan on the spongy tundra and the rosy-finch on the overhead peaks, and I'd failed to find either bird despite making five trips to their alpine habitat across the previous two weeks. Those iterated reaches required a full day of climbing split between biking and hiking, and each unsuccessful attempt left me more exhausted than the previous. Colorado's Rockies were kicking my ass, and I didn't have the energy to yo-yo indefinitely.

Bedding down in Georgetown on June 26th, Guanella Pass would be my final opportunity to find the ptarmigan and the rosy-finch along my envisioned route. If I missed the birds there, then I'd need to accept delays or make significant detours, which would erode my time buffer and further exhaust me. I wanted to find the birds and avenge my previous misses, but I needed to avoid overexerting myself in the process. Risking injury or burnout wasn't prudent with six months of riding still ahead of me, but the thought of finishing the year with 598 species absent the alpine pair was unbearable. Fearing that outcome more than injury, I readied for my Guanella assault.

SIXTEEN

Mountains and Valleys

My Guanella Pass (11,670-foot) ascent, slated for June 27th and seeking White-tailed Ptarmigan and Brown-capped Rosy-Finch, would be the steepest 11 miles I'd faced among the 7,800 I'd covered to Georgetown, Colorado. The 3,100-foot climb would be challenging under any circumstance, but exhaustion from a crippling 3,500-foot climb over Loveland Pass (11,991 feet) the previous day, an ordeal I barely survived after a day off, guaranteed Guanella would be grueling.

Afternoon thunderstorms would complicate my undertaking by limiting the time I could spend above the tree line. The alpine tundra is about the worst place on Earth to be caught in a lightning storm, so I would need to summit the pass, scour the exposed habitat for the ptarmigan and rosy-finch, and evacuate ahead of electrical discharge. If I struggled on the climb, then I'd be forced to abort and wait for more favorable conditions. The safe play would have been to delay while I rested and awaited better weather, but the Big Year clock was ticking, days becoming more valuable as the year's midpoint approached.*

I left my motel and confronted a steep set of switchbacks where Guanella Pass Road departed town. My fatigued hamstrings and weary

---

* Ten tundra hikers would be hospitalized and two killed by lightning strikes in nearby Rocky Mountain National Park two weeks later, on July 12th and 13th.

quadriceps contracted with the fortitude of old underwear elastic, and my hypoxic calves tightened as I pushed through the punishing hairpins. I overcame several cramps without dismounting, but a violent seizure gripped my right calf on the steepest pitch; it felt like someone was driving a nail through my lower leg. Unable to unclip my foot before the murderous grade brought me to standstill, I flopped helplessly right, my bicycle crushing my cramped leg against the tarmac. I screamed. I flailed. I fought to free my paralyzed limb from the pedal. That task accomplished with the aid of my arms, I stood, the ensuing two minutes excruciating as I overcame the contraction with painful stretching. The weather forcing me forward, I took a few test strides before remounting and continuing.

Riding with unclipped feet, I staved off additional cramps before they grounded me again. The switchbacks abated, the road assumed a saner incline, and I settled into a steady rhythm as the mountain thoroughfare skirted a series of lakes and wound higher into the Arapaho National Forest. Pine scent tickling my nose and birdsong tap-dancing on my tympanum, I made halting progress toward the tundra.

An upper collection of switchbacks renewed my misery. Gravity dragged on my bicycle like a ten-ton anchor, and I thought my chest might explode as my heaving lungs tried to supply oxygen to my throbbing legs. Each pedal crank more contentious than the previous, I oozed skyward.

A snow-covered Mount Bierstadt (14,065 feet) came into view; guarding the eastern side of the pass, it stood protectively, almost paternally, over the supporting tundra. A final push powered me into the parking lot at the saddle, and a spectacular collection of frosty peaks stole what little of my breath remained. I knew I wasn't on top of the world, but my eyes were telling me otherwise.

I couldn't afford to linger. I broke the landscape's spell, swapped biking shoes for hiking boots, and stashed my bicycle behind a stand of bushes. Following a rudimentary trail across the spongy tundra, I gained a small ridge before dropping into an expansive bowl, the perfect habitat for ptarmigans.

Ptarmigans are pigeon-sized grouse that inhabit alpine and Arctic tundra. Patches of low bushes are the only cover in those treeless

expanses, so the birds have evolved seasonal camouflage to evade preda-
tion when venturing beyond those punctuated protections. In summer,
their mottled gray-brown garb mimics the lichen-covered rocks that dot
their tundra digs; in winter, their white costumes render them ghosts
against drifting snow. Ptarmigans hide in plain sight, the lethargic birds
running or flying only when predators are on top of them. Searching for
ptarmigans is a brute force function; the more tundra one explores, the
better the probability of intersecting a representative.

Thinning air left me gasping as I moved above 12,000 feet. Assess-
ing gathering clouds while I paused to recover, I gauged I could invest
another hour before initiating my retreat. As I caught my breath and
rejoined the path, a football-sized rock rolled across the worn earth five
feet ahead of me.

What the hell?

Disregarding gravity, the earthen blob shuffled onto the uphill side of
the path. I stumbled backward in disbelief but pulled focus to see a female
ptarmigan staring at me from ten feet away. I reached for my binoculars.

She's too close for those! Use the camera!

I reached for the device, but its shoulder strap had inter-looped with
my binoculars during my retreat. Frantic, I tried to untangle my optics.

*Hurry! Before the damn thing flies away! These pictures could be awesome
in the blog!*

Another twenty seconds elapsed before I freed the camera and
dialed in the appropriate settings, but the bird didn't move an inch, my
fumbling suggesting to the ptarmigan that I was more dolt than danger.
Apparently reassured by my ineptitude, she shuffled off the path and
onto some spongey foliage as I trailed closely behind. Her concealing
plumage was beautiful, but I found her feathery legs and feet even more
captivating. Accented with scarlet eyebrows, she walked the tundra catwalk with quiet confidence. She was remarkably trusting, and I savored
the encounter against my inability to intersect the bird on five previous
reaches to the alpine zone. It was the most I'd invested into a single
species to that point in my journey.

Thunder audible toward the end of my allotted hour, I left the ptarmigan and started my retreat. Scampering down the mountainside, I

marveled at her durability. I thought I was tough for summiting Guanella on a bicycle, but the ptarmigan forced a humbling recalibration; where I struggled to survive, the unbreakable bird thrived.

The ptarmigan was a huge victory, but I failed to find my other target, Brown-capped Rosy-Finch. Fortunately, I had an insurance policy in the form of Chris Rurik, a local birder and blog reader with whom I'd been communicating. A recent Denver transplant, Chris hadn't found the time to seek the rosy-finch since moving from Seattle a year prior. Making my urgency his own, he'd volunteered to help me at Guanella Pass Campground that evening. The primitive overnight site is only 900 feet below the saddle, so his overture—complete with tent, sleeping bags, and food—would allow me a second Guanella day without dropping all the way to Georgetown for the intervening night. Prior commitments rendered his offer good on Friday night only, and it was that consideration which motivated me to attack Guanella without delay.

I spent the remainder of Friday afternoon dodging raindrops below the tree line and met Chris at the campground in the early evening. In his mid-twenties, he was tall, lean, and neatly bearded. We set up the tent and prepared dinner, and a previously undiscovered Stanford connection fueled nostalgia during our after-dinner walk. Heavy rain forced us into the tent at nine p.m., and several waves of violent thunderstorms went through overnight—at least according to Chris. I was so tired I slept through them.

The storms cleared by morning and left blue skies behind. I departed the campsite ahead of Chris, and we rendezvoused in the summit parking lot forty minutes later, our plan to make a direct line for Mount Bierstadt. Unlike the ptarmigan, the sparrow-sized rosy-finches are highly nomadic. They move from peak to peak throughout the day, so the best way to intersect them is to obtain a high vantage and wait for the brown-and-pink birds to fly in. Birders generally find ptarmigans, but rosy-finches often find birders.

Weaving around weekend hikers, Chris and I gained elevation. Our ascent was slowed by ice on Bierstadt's upper throws, the overnight rain to blame, and I weighed the risks of continuing. My health hung on each slippery step, so I was keen on caution. I pointed to an overhead ridge. "That spot looks perfect," I said. " It has a good view of the upper slopes, so we don't need to go to the top."

Chris countered, "Yeah, it's OK. But we should go for the summit. The view will be awesome."

It was our first strategic disagreement. Chris was as intent on the summit as I was on the rosy-finch. The goals weren't mutually exclusive, but I didn't want to expend more energy than was necessary to find the birds.

I clarified my position, "I'm worn out from the last few days. I don't want to fall on my ass or bust my leg. I've got another six months of this shit."

Chris replied, "The summit isn't much higher. You've biked and hiked a million miles already. You'll be fine."

As I thought how to reconcile our priorities, a pudgy beagle scrambled up the icy trail and pawed past me; his owner could barely keep pace with the enthusiastic hound. My ego clicked on; there was no way I was going to be outdone by that kibble-fed bowling ball. If the beagle was going to give the summit his best effort, then so would I. I caught up to Chris, and the two of us followed Snoopy up the mountainside.

Just below the summit, we encountered a rocky chute passable by only one hiker at a time. While we waited our turn, Chris wandered over to the edge of a precipitous drop. Hoping to avoid a plunging death, I stayed put and held our place in line.

Chris motioned to me two minutes later. "Come over here," he said calmly.

I begrudgingly complied, surrendered our place in line, and walked thirty yards to join him. "What's up?"

He pointed down the cliff face. "Look. Way down there," he said.

I didn't see anything except for scree and boulders, but a twitch caught my eye. I raised my binoculars for a better look. I exclaimed, "Brown-capped Rosy-Finches!"

The four birds rocketed up the cliff face. Three flew directly over us, but the straggling fourth landed thirty feet to our right, its milk-chocolate body and dark-chocolate cap contrasting with its pink belly and flanks as it posed for pictures. The bird departed twenty seconds later, and Chris and I celebrated Big Year bird #488 with fist pumps and high fives.

Buoyed by adrenaline, we bounded through the bottleneck and reached the summit. The view from 14,065 feet was breathtaking, with Mount Evans (14,265 feet) visible to the east, and Chris and I borrowed a cardboard sign from another pair of hikers—"Mt Bierstadt 14,065 feet"—and snapped a photo with it. The beagle survived the climb, too, and he offered a curious sniff and an approving tail wag as we lounged on the sun-drenched summit through the next half-hour.

Our descent was a sobering reminder of the need for caution. We saw several people fall, and one older man was awaiting evacuation on account of an injured ankle. Reaching the parking lot, I thanked Chris for his help and company. Without his motivation and assistance, I might have missed the rosy-finch and failed to experience the summit. My alpine business handled, we parted ways.

From Guanella Pass, I had eleven downhill miles to return to Georgetown and another thirty to reach my Genessee overnight. Mountains and valleys necessarily experienced proportionately, I was thankful for gravity's assistance as I rocketed out of the Front Range.

———

Walking into the basement of the Midtown church in December 2006, I thought my life was over. The white walls mirrored my blank vision of the future, and a flickering light suggested sobriety would be a dim journey. For ten years, since my junior year of high school, every fun night, every story worth telling, and every sexual encounter had been facilitated by alcohol. Sobriety would never replicate the instant gratification and disinhibited euphoria of drinking and drugging, and I expected my future would be a depressing slog of predictability, isolation, and masturbation without alcohol and narcotics.

Despite the promised anonymity, there was nowhere to hide among seven other attendees at that first AA meeting, my trepidation heightened by our arrangement around a circular table. The gathering was called to order, and the Drunks of the Round Table introduced themselves and declared their current sobrieties. Unable to hold back tears when my turn came, I sobbed the dreaded utterance, "My name is Dorian. I'm an alcoholic. This is Day One."

Soft applause didn't encourage me, and I fell silent for the remainder of the meeting, my teary gaze split between the floor and the clock for fifty-nine minutes.

That first meeting was awful. Proclaiming my alcoholism to a group of strangers was a humiliating defeat and a stinging admission that alcohol and drugs had abused me even more than I had them. Despite continued feelings of shame, failure, and inadequacy, I dragged myself to the church basement each evening for the following week. I didn't contribute much to the discussion, but my time in the room revealed that each person had a unique story punctuated with the same terminal mark: a crippling inability to regulate alcohol intake. Moderation impossible, surrender emerged as the only option, often after tragedy. It was in the collective recognition of shared desperation that AA hoped to cultivate acceptance, healing, and lasting sobriety.

My biggest hurdle to achieving those goals was my identity; I wanted to get sober but not more than I wanted to admit the fallibility of the persona I'd maintained across the previous decade. Sobriety felt like quitting myself, and I dreaded reinventing myself without familiar crutches. As I told my cousin, Alexis, alcohol and drugs would still be there if I decided sobriety sucked.

A week into the program, the intimacy of the church gathering was suffocating me, so I switched to a larger group—the Mustard Seed at 37th Street and Lexington Avenue—to stay engaged. Fifteen walking minutes from my lab at the NYU medical center, the new group invigorated my commitment; with twenty folks from a rotating bevy of forty attending the six thirty p.m. meeting, presence and personality were in ample supply. I was comfortable sharing my thoughts on most evenings, but the group was large enough that I could hide in the back when I didn't feel like participating.

I established a daily routine of lab work, my AA meeting, and alone time at home. The December days were tough, bars and nightclubs calling to me from the long winter darkness, but I extended my sobriety against those nocturnal whispers. I hadn't experienced a week of sobriety since leaving the restrictive confines of Hotchkiss nine years earlier, so twenty days in the program represented a significant accomplishment.

My routine was insulating and protective, and I was optimistic I could achieve ninety AA meetings in ninety days, a suggested goal for newbies.

My fellow attendees were friendly and entertaining. I appreciated their honesty, valued their support, and recognized the commonalities that brought us together. Despite those connections, I drew distinctions between their experiences and my own. Many had lost jobs, been ostracized by friends and family, or experienced financial ruin, and I believed my social and celebratory approach to alcohol and drugs was different from those who confessed to drinking and using—often alone—out of frustration, depression, or anger.

Despite the refrain I uttered at the beginning of each AA meeting, I didn't think I was a *real* alcoholic. I just liked to party; that I got dangerously drunk and high multiple times each week didn't matter. I didn't find the program's foundational Twelve Steps helpful, especially those predicated on the acceptance of a Higher Power, and I didn't make an effort to connect with individuals in my home group, using meetings more as ritual than opportunity. One guy, Albert, approached me after I'd been in the program for a month. He helped organize and run the meetings, a cup of coffee glued to his right hand throughout, and I thought of him as a positive presence.

"Hey, man. I've been enjoying your shares these last few weeks," he said.

"Thanks. I'm still getting the hang of this whole thing," I replied.

He renewed his current chemical romance, lowered his cup of coffee, and spoke, "Yeah, we've all been there. A sponsor might help. You found one yet?"

A sponsor is a veteran member of the group who offers personalized guidance. There are no codified rules, but it's recommended that a person have a year or two of sobriety before extending sponsorship offers.

I replied, "No."

"Well, don't wait too long. Having someone to talk to is a big help," he said. "I sponsor a couple of folks but have room for another if you're interested."

Albert claimed a decade of sobriety, and his nightly shares revealed he was friendly and honest. I replied, "That's great, thanks. But I'm still getting to know everyone here. I want to give it a bit longer before committing. I'll keep you in mind. Thanks."

Albert exhaled deeply and gave me a look as though he'd heard that deflection a thousand times; we both knew I didn't want the additional accountability a sponsor would provide.

He spoke with lowered expectations, "Well, I'm available when and if you're ready to make the commitment."

Beyond that excuse, I didn't tell anyone outside the program about my sobriety, Alexis included, because I felt damaged and inadequate. When I should have enlisted family, friends, and colleagues as support, I kept my situation secret so that I wouldn't look like a failure or let anyone down if I relapsed. Apart from my nightly AA meeting, I didn't take my sobriety out for exercise that first month, keeping it locked away from view for the remaining twenty-three hours of each day.

Instead of embracing sobriety's obvious benefits—physical health, emotional stability, professional productivity—I dwelled on its imagined costs, specifically the death of my late-night lifestyle. Worse, I failed to formulate a sustainable view of the future because I viewed sobriety as penance for past behavior. Like a cramped calf muscle, hindsight and regret prevented me from overcoming the steep but surmountable hill in front of me.

Together, those attitudes suggested my sobriety was a fling rather than a relationship. I viewed the suggested "ninety-in-ninety" plan as a time to complete my sobriety rather than begin it, and I convinced myself I could approach drinking differently if I survived that three-month crash course. Extending that painfully alcoholic line of reasoning, I surmised thirty-eight sober days were as good as ninety, and I picked up a forty-ounce bottle of Olde English malt liquor on the walk home from my final AA meeting. Waking in a soaked bed and with no memory of what transpired in the intervening hours, I still didn't understand that I was exactly the same as everyone in the program I'd just abandoned.

SEVENTEEN

Chickening Out

The prairie was scorching, the sun's rays prodding at my leathery skin like a barbecue master poking at a slab of sizzling meat. Knee-high grasses extended in every direction, and a slight breeze bent the wispy blades in hypnotic unison as I rolled east on Highway 14. I couldn't believe I was only a week removed from Bierstadt's summit. Since that icy pinnacle, I'd surrendered 10,000 feet of elevation to reach Pawnee National Grassland east of Fort Collins, Colorado. Giddy on the extra oxygen at that lower elevation, I powered into the heart of the 200,000-acre preserve.

Pawnee was a scheduled deviation from the mountains to search for a handful of prairie birds before I tackled more extreme topography through Wyoming, Utah, Idaho, and Oregon. Administered by the United States Forest Service, Pawnee is designated as "multiple use." That classification permits economic activities on federal land, and ranching and resource extraction have irrevocably fragmented what was an unbroken carpet of short-grass prairie in pre-Columbian times. There is still much intact habitat, but manifest destiny and permissive policies have guaranteed neither Native Americans nor American bison would grace the rolling landscape as they once did.

Thankfully, birds have persisted in healthy numbers, and Lark Buntings sang from every fencepost as I wheeled into Pawnee midday on July 3rd. Colorado's state bird and a member of the sparrow family, the male is glossy black with bold white wing patches; the female is brown above and creamy below. Their song is a rapid progression of whistles and trills and carries a great distance over the sun-soaked prairie, the chorus deafening as I explored Pawnee's dirt tracks. Culling through the melodious hoards that afternoon and the next day, I teased out McCown's and Chestnut-collared Longspurs, sparrow variants named for their elongated hind claws, and Mountain Plover, an upland shorebird, to push my total to 495 species. With my prairie business concluded on the evening of July 4th, I contemplated my next move.

Up to that point, I hadn't significantly deviated from the ideal route I imagined from Massachusetts. I assumed inexperience, illness, and injury would force me to compromise aspects of that ambitious 15,000-mile itinerary, but strengthening legs, extended health, and timely birding had me running two weeks ahead of schedule at year's midpoint. Extra time in hand, I considered expanding my route to search for additional species.

Greater Prairie-Chicken was one such possibility. The species historically ranged throughout the Midwest, but the unyielding conversion of long-grass prairie into farmland has forced most of the remaining birds into isolated pockets in the Dakotas, Nebraska, and Kansas. I excluded the bird from my original plan because I didn't think I'd have time or energy to visit those states, but—in reevaluating the species from my accelerated perspective—I learned it also ranged into eastern Colorado, a geography I could reach before proceeding to Wyoming. The round-trip detour to Wray, Colorado, a stronghold for the declining bird, would require a minimum of 300 miles and deplete my two-week buffer by at least five days. With no guarantee I'd find the shy bird, the undertaking would be a huge gamble.

I'd designed my route so that I'd exhaust the expected species exactly at the end of the year; finishing too fast, with extra days in hand, would be a waste. If I'd been more confident in my cycling abilities earlier in the year, then I would not have departed the Texas coast without additional searches for Black-billed Cuckoo, Mourning Warbler, and Yellow-bellied Flycatcher. I didn't regret that decision because I acted to preserve my health and conserve time, but I'd clearly underestimated my cycling ability. That realization suggested I should be more aggressive in the second half of the year than I'd been in the first.

My ambition swelling, I departed Pawnee on July 5th and headed east, toward Sterling. Northeastern Colorado is sparsely populated, and feelings of isolation renewed in those undeveloped surroundings. Loneliness was, as it had been previously, compounded by my decision to forgo listening to music, audiobooks, talk radio, or anything else while riding. Pragmatic concerns of hearing singing birds and approaching vehicles aside, I wanted to spend every riding minute with my own thoughts. That decision facilitated introspection on some days but drove me batshit crazy on others, the balance tipping toward the latter as I cranked toward Holyoke on July 6th.

Starved for interaction, I sarcastically mooed at a herd of roadside cows. When they responded, I increased my mimicry and thereby drew more bovines into the impromptu choir. Our volley intensified as I continued along the pasture's edge, and a group of twenty broke from the herd and paralleled me on the other side of the fence that separated us. When I accelerated, the animals did the same, cantering behind their two-wheeled Pied Piper. It was an improbable connection that ended only when a fit of laughter forced me to stop and catch my breath. I was unable to inspire a second pursuit after collecting myself, but I extended the game to additional cows through the afternoon. With sheep and pigs also in the mix, I was mooing, baaing, and oinking my way across Colorado. Some solitude is healthy; too much is maddening. But with the livestock listening, I did feel less alone.

I approached Wray in the late afternoon and commenced my prairie-chicken search north of town. Reminiscent of domestic chickens at first glance, the brown-and-tan birds avoid detection by slinking through waist-high scrub. My only hope was to flush a representative from the concealing vegetation, so I increased my presence by clapping and yelling as I high-stepped through the habitat. I was thankful there was no one to witness my preposterous performance; I suspect the folks in white coats would have been called otherwise.

Despite stomping around for four hours in ninety-five degree heat, I was unable to find the bird on that first afternoon. A five-hour attempt the following morning proved similarly chicken-less, and I faced a no-win decision as my July 7th afternoon outing drew to a close: backtrack to Pawnee without the bird or push additional days into the search. The first would be a demoralizing defeat, but the second guaranteed nothing beyond more time invested.

Weighing my options as the sun sank, I felt my heart stop when I heard what sounded like someone starting a lawnmower fifty feet to my right. Spinning in that direction, I saw a Greater Prairie-Chicken taking flight, its broad wings the source of my startle as the heavy-bodied form labored into the air. The bird and I locked eyes for a split second, and the escapee fled over a low rise to the west. I sprinted after it, hoping for another glimpse before it regained cover.

The bird was gone when I reached the hill crest, but I hardly cared. A fuchsia orb backlit every blade of grass and tuft of vegetation, and a fine, airborne particulate created a peach-pink shroud as the sun's final rays diffracted through it. With hundreds of glowing insects floating through the prairie panorama, the picture was perfect; any additional intrusion on my part would ruin it. Solitude suddenly more an asset than a liability, I stood transfixed, the diurnal warbles of Western Mead-owlarks yielding to the nocturnal buzzes of Common Nighthawks as the sun retired.

It was dark by the time I returned to my bicycle and regained Highway 385. Fifteen miles between me and my Wray motel, I flipped on my light and headed toward town for a second night. The heavens consumed the prairie, and I felt an insignificant actor as I pedaled across the planetarium stage.

———

I departed Alcoholics Anonymous in January of 2007, after thirty-eight days of sobriety, and immediately resumed the drinking and drugging patterns that drove me to the program. While AA failed to convince me that I was an alcoholic, it did suggest I had a drinking problem. As I believed all problems were at least partially solvable by my analytical mind, I implemented a series of regulatory strategies to manage my condition.

My attempts to confine drinking to weekends failed miserably, and a subsequent effort to limit consumption to six drinks per outing also proved futile. I tried allotting myself $25 per night, but that was pointless because compromised judgment inevitably rendered limits ignored. My boldest stroke was cutting up my ATM card because I thought restricting access to late-night cash would reign in renewed cocaine use. It didn't, nor did it address the underlying drinking from which the drugging stemmed. No matter what roadblocks I implemented, I was unable to attenuate intake; alcohol and drugs were enjoyed in direct proportion to their abuse.

Incongruously, my scientific research proceeded at a fantastic pace during my renewed substance abuse, and I was selected to present my work at an international conference in Los Angeles in June of 2007. My plenary talk was applauded by thousands of scientists, several Nobel laureates among them, and I celebrated at the barbecue that followed the session. Instead of networking, I drank, my intoxication extending to a bar in downtown Westwood afterward. Scoring acid at that dive, I spent the remainder of night in the sexual company of the woman who facilitated my spontaneous psychedelia. By my measures, it was an average Thursday, transient interactions with drugs and drug users prioritized over lasting connections with friends and colleagues.

That trend crescendoed two weeks later, when I traveled to Hotchkiss for my tenth-year reunion. While my classmates enjoyed a cocktail or two, I went on an Adderall-fueled mission to drink the campus dry. I blacked-out for several stretches, and a few of my classmates had seen enough by the end of the weekend. James, one of my two best friends through Hotchkiss, pulled me aside as I headed toward the train station on Sunday morning.

"Look, man, I want to talk to you about my wedding," he said.

"Yeah, looking forward to it. Should be a helluva party," I replied.

"Well, that's the problem," he said. "It's a big day for Erin and me, and I gotta know you aren't going to be a distraction."

I understood his insinuation. "So, what? You want to put a leash on me?" I joked.

"No, I want you to get your shit together," he said forcefully. "You were hammered all weekend, and I don't want a repeat performance in October."

I didn't appreciate the judgmental tone. I snapped at him. "When did you become the fun police? Fucking A, James. I had my first beer at your house. You invited me to DJ at Williams when you were a student there. You and I threw a massive kegger at this same reunion five years ago, and . . ."

"And you got so drunk that you pissed on the floor!" he shouted. "I figured this shit would've ended by now, but you're worse than ever. I mean, how the fuck do you get anything done at NYU?"

"I handle my business and then I party. What's the fucking problem?"

"You're a fucking liability!" he shouted as he repeatedly clapped the back of his right hand against his left palm. "I'm tired of qualifying you to people. I'm tired of worrying if you're going to do something crazy. And I'm tired of apologizing when you do. Your drinking is demanding headspace I don't have."

The exchange building, I asked the relevant question with my arms thrown out. "And the wedding?"

"You've made this really easy. I don't want you there."

Those would be the last words we'd speak for five years.

Several other friends subsequently excluded me from their weddings, but I hardly cared. I had a fresh crop of folks in New York with whom I'd rather spend my time. Aided by those enablers, my drinking and drugging reached new heights during the second half of 2007, when ketamine and methamphetamine joined the narcotic rotation. The former offered dissociative, out-of-body experiences while the latter energized me, but I didn't particularly like either compared to ecstasy or cocaine. Unable to refuse anything once I was annihilated, I used them only because they were there, on offer from others, at all-night loft parties.

There was periodic friction with Jeremy, my advisor, through that year, but we'd reached an equilibrium which was mostly "don't ask, don't tell." As long as exciting data continued to pour in, I knew I had a scientific future. That sentiment peaked when I published a lead-author paper in the prestigious *Science* magazine a year later, in June of 2008. That achievement guaranteed postdoctoral opportunities after graduate school and affirmed for me—however falsely—that academia would tolerate my alcoholism as long as I could produce interesting data. Through warped lenses, I saw myself drinking and drugging under the guise of scientific productivity well into my thirties. Compared to my stint in AA eighteen months earlier, that imagination represented a catastrophic devolution.

Though it was rampant, my substance abuse hadn't imparted enough negative consequences to motivate change. Hangovers passed, wet beds dried, cuts and bruises healed, and the concerns of lifelong friends became harder to hear against the encouragement of enablers and the din of dance music, my DJ side gig and clubland adventures still going strong. I didn't understand that each drink exposed me to more risk than the previous or realize the degree to which my choices had become unsustainable. Approaching thirty years in age, I didn't think the physical or emotional bills for my choices would ever come due. Blinders in place, I staggered into alcoholic darkness as the fall of 2008 arrived and my fifth graduate year began.

My reminiscence faded as I reached Wray's illuminated outskirts. Revisiting my recollection after showering, I couldn't believe how my thinking had evolved since my drinking days. While alcohol was reflexive and precluded considerations of negative outcomes before plunging headlong into the next blackout, the bicycle was deliberate and required premeditated cost–benefit analysis at every juncture. Most of what I did in my twenties was done in the moment, but the bicycle forced consideration because of the energetic and emotional outlay it required. Thankful that my Wray detour had highlighted that maturation, I passed out, images of prairie-chickens dominating a deep and restful slumber.

Waking early the next morning, the 8th, I retraced my tracks to Holyoke before returning to Sterling. Rather than backtrack through familiar parts of Pawnee on the 9th, I cut north into the preserve's eastern unit with hopes of adding Sharp-tailed Grouse en route to Pine Bluffs, Wyoming.

An intimidating wall of thunderheads materialized in the afternoon, and I abandoned my grouse plans ahead of approaching danger. The wind increased, the temperature dropped twenty degrees, and lightning struck as the deluge intensified. A bolt grounded 200 yards from me; zero delay between flash and crash, an earth-shaking rumble spread across the prairie. My spine tingled, terror gripping me as electricity attacked the landscape.

I was the tallest entity for miles. I knew my best chance to avoid a strike was to lie face down on the ground until the storm passed, but cowering was not my preferred recourse. I kept on the pedals, visibility decreasing to near-zero as I battled torrents and gales. Dirt tracks muddied, and I dismounted and pushed through several impassable sections. I pressed north, the thunder and lightning unrelenting.

The storm cleared to the south through the ensuing hour, but resulting headwinds made the final twenty miles to Pine Bluffs miserable, particularly as I was soaked from head to toe. By the time I reached my motel, I'd covered ninety-five miles. Those miles pushed the week's total to 560 and left me feeling incapacitated. My prairie-chicken victory forgotten, I wondered how the hell I would survive 400 miles of blustery desolation as I crossed southern Wyoming on Interstate 80 in the next week. Just two days after feeling invincible, another six months on the bike seemed impossible.

The next day, July 10th, I limped fifty miles to Cheyenne, where I'd connected with a host family. Ed, my online contact, wouldn't be home but had arranged for his wife—Ruth—and kids—Sally and Josh—to receive me. Ruth was at work when I arrived, so the two teenagers ushered me to the guest room of the two-story house. I took a shower and joined the kids in the living room.

"So, what do I need to know about Wyoming?" I asked the pair as I plopped into a leather recliner.

They looked at me silently.

"OK, let's start smaller. How long have you lived in Cheyenne?" I asked.

"Like, forever," Sally said.

"Really? Forever? Since dinosaurs roamed the Earth?" I asked sarcastically.

Josh snickered while Sally clarified, "We were both born here if that's what you mean."

"Yeah, that's what I mean."

I chipped away at their adolescent indifference, and complete sentences slowly prevailed. Sally, a rising senior, elaborated on her college plans. Josh, a soon-to-be freshman, avoided eye contact and spoke shyly about baseball from under the brim of his threadbare cap. Both were friendly and funny, and their company was a welcome reprieve from the week's isolation.

Ruth returned home in the early evening and asked countless questions about my adventure while she prepared dinner. Sensing my curiosity as I watched Sally and Josh exercise a pair of cows in the backyard, she encouraged me to join them. She taunted, "C'mon, city slicker! Let's see what you've got!"

I grabbed my hiking boots and headed out back.

Sally instructed, "Take this halter and walk slowly around the pen. Bessie will follow."

I took the rope, but the beast wouldn't budge. Pleading was ineffective, and I slipped into the mud when I tried to drag the obstinate heifer forward.

Actively leading the second cow, Josh suggested, "Let's switch. Minnie is more cooperative."

I'd barely surrendered Bessie's halter when the stubborn animal kicked into gear. Minnie correspondingly stopped when I reached for her rope and wouldn't reengage no matter what I did. Pushing forcefully on her rear end, I played it up for the teens. "Moo-moo-move," I said. "I want to introduce you to my friend, the Burger King!"

The teens laughed as I leaned on the immovable bovine, her resistance likely payback for the shameful number of hamburgers I'd consumed in my travels. I conceded defeat, the cow poop on my boots a reminder of my cowboy shortcomings.

Dinner was delicious. With wind and rain forecasted for the following day, Ruth insisted I forgo riding and stay a second night, an offer I accepted without hesitation. I needed the rest, and another night would allow me to meet Ed when he returned from his business trip.

I spent the entire following day in front of the television. Ed appeared in the afternoon; offering snacks and conversation, he was as cheerful in person as he was on the phone. Ruth threw together another elaborate feast, and a dozen of us—the four family members, me, and seven neighborhood friends—sat down to break bread.

One particular exchange stood out at that gathering. From what I could piece together, Ed had tasked Josh to put up a fence along a stretch of their property, a chore Ed wanted done by the time he returned from his trip. The fence was complete but crooked, so Ed suspected that Josh and Mark, Josh's assisting friend, might have been drinking when they erected it.

Ed voiced his suspicions at the dinner table. "I'm not going to be mad," he said. "I just want you to tell me the truth. Did you and Mark have a few beers before you put up the fence?"

Speaking through a mouthful of mashed potatoes, Josh replied, "No."

Ed put down his fork and tapped the table with his finger. "Are you sure?" he asked.

Josh seemed to be enjoying the joust. "Yes," he said after a brief pause, a pea falling out of his mouth during his declaration.

Ed was trying not to laugh for fear of tipping the contest in his son's favor. He collected himself and pleaded, "It's OK. You can be honest with me."

The reiterated plea caught Josh between mouthfuls and unable to stall. "Well, um, Mark mighta had a few beers before," he said. "But I dunno for sure."

Everyone at the table knew Josh wouldn't give it up, but Ed thought he saw a crack in his son's position. Again holding back laughter, he calmly asked the pivotal question, "And you? Did you have any beers?"

"I can't remember. I was focused on the fence!"

"Sure you were," Ed said with an eye roll and a smile.

Everyone at the table was cracking up, and nothing was resolved save for the fence being crooked. After a solitary week, it was wonderful to have such warm company for those two nights. Instead of a singular host, I'd found family. They were exactly what I needed after a long and lonely prairie-chicken chase.

EIGHTEEN

Secrets of the Sage

Reevaluating my route while I rested in Cheyenne, I realized my plan to cross southern Wyoming was likely suicidal. My legs were pulverized after the prairie-chicken detour, and far-flung stopovers along Interstate 80 would require energy I didn't have. Rather than endure a week of potentially crippling western headwinds along that corridor, I decided to ride to Laramie as planned but tack south into northern Colorado from there. That alternative would shelter me from Wyoming's notorious gales, offer more civilization and support, and utilize quieter state highways. It would also afford an opportunity to find my next target bird, the beautiful and bizarre Greater Sage-Grouse.

Thirty inches long and occupying the volume of two misshapen basketballs, the Greater Sage-Grouse is the largest native game bird in North America. Like their prairie-chicken relatives, these gray-and-brown ground-dwellers are shy and lethargic, ranging through Colorado, Utah, Wyoming, Montana, Idaho, Washington, Oregon, Nevada, and California. Though the birds likely numbered more than fifteen million in the early 1900s, their population plummeted to an estimated 350,000 by 2014—a decline of 97.5 percent—as housing, mining, ranching, and fossil fuel extraction fragmented and degraded their obligate sagebrush habitat.

Surveys suggest two-thirds of the remaining population reside on public lands, so the species has emerged as a political football in the western United States; with much state and federal acreage designated as "multiple use," the bird's survival is often pitted against economic motivations. The Bureau of Land Management and the United States Forest Service have tried to balance conservation concerns against industry advocacy, but the agencies have been paralyzed by political indecision. Without a viable policy in place, private industry continues to exploit public lands, an amorphous situation leaving the Greater Sage-Grouse in continued peril.

Circumventing bureaucratic inertia, private landowners have emerged as unlikely leaders of Greater Sage-Grouse conservation. If sage-grouse populations fall below a sustainable threshold, then the federal government will list the birds as threatened or endangered; those designations would grant regulatory agencies increased influence over whatever private lands harbor the birds. Landowners are loath to cede control of their acres to the federal government, so some have voluntarily curtailed economic development on their properties because predictable, short-term sacrifices are preferable to longer-term losses if the feds were to intervene in the future. The interests of the birds and private landowners are oddly coincident, and voluntary self-regulation is an encouraging example of citizens considering wildlife in their economic equations, even if motivated by underlying economic self-interest.

Independent of conservation status and political maneuvering, the Greater Sage-Grouse is one of the most magnificent birds on the continent, a claim that manifests during the spring mating season. Greater Sage-Grouse do not mate in a pairwise manner as do many other birds. Instead, they utilize a polygamous system that allows the most fit males to mate with many females. The centerpiece of their unusual courtship is the lek, a historical gathering ground where dozens to hundreds of birds assemble each spring. Biologists don't know how leks are designated, each just a grassy rise or patch of worn earth in the surrounding sagebrush, but the arenas facilitate the most incredible mating displays on the continent.

After commuting to the lek from his individual territory, each male stands erect, expands his barrel chest, and fluffs an extravagant white ruff from his neck. Like a flamboyant drag queen in his fluffy boa, he inflates two yellow air sacks in his chest, those orbs jiggling and bouncing like bikini-clad breasts as he cyclically builds and releases the pressure in them. He holds his wings slightly open at his sides as though inviting conflict—"come at me, bro"—and he splays his tail in a semicircular halo recalling the Statue of Liberty's thorny crown. The original transformer, no amount of computer-generated imagery could replicate his vitality.

The inflated males strut around the lek and deploy pecks and kicks as they bump into each other and squabble for superiority. The spectating hens interpret the machismo, and the majority mate with the handful of males they judge exceptional. Each lek operates for several weeks in spring, and then both males and females disperse into the habitat as hormones equilibrate and mating ends.

The male's contribution to the next generation is purely genetic; he disappears into the sagebrush to await next year's courtship while the female lays the eggs, incubates the clutch, and raises the young. Though the male is an apparent deadbeat dad, his absence is evolutionarily forgivable because grouse chicks are very precocious. The female need not feed nest-bound young as songbirds and raptors do; instead, she leads the ambulatory hatchlings to stream-fed meadows where they gorge on succulents and insects. This rearing system allows grouse clutches to be among the largest in the avian world, and it permits the most-fit males to sire many more offspring than if they were forced to parent a single brood full-time. The downside of the mating system is that in theory it should collapse genetic diversity—insurance against shifting ecological pressures—to a female-dictated mean. How Greater Sage-Grouse and other lekking species maintain genetic diversity using a mating system that should reduce it is a beautiful mystery.

Given the pomp and circumstance surrounding mating season, March and April are the best months to seek Greater Sage-Grouse and the other lekking game birds (such as Greater Prairie-Chicken and Sharp-tailed Grouse). Big Year birders typically fly into Denver and drive a foolproof loop through Colorado and Kansas, which nets them all the relevant birds in a week. I needed to be on the Gulf Coast to pursue transient migrants at that season, so my searches were delayed until summer, a season when the birds are dispersed and very difficult to find.

I began my sage-grouse search on July 13th, at the Arapaho National Wildlife Refuge south of Walden, Colorado. In the center of a

400-square mile basin created by a ring of snow-capped peaks holding back the outside world, I toured the gravel auto loop as it wound between sagebrush and wetlands. Ducks and shorebirds filled the impoundments, eagles and falcons soared overhead, and I imagined the scene representative of centuries prior—minus a lack of sage-grouse. The experience was enough, and only afterward did I realize that Eared Grebe, a duck-like bird with golden tufts on its head, and California Gull, a geographic misnomer for a bird ranging across the American West, had pushed my total to 500 species. At that moment, it seemed the best way to achieve my 600-species goal was to temporarily forget about it.

I departed Walden the following morning and resumed my search on back roads en route to Steamboat Springs. My eyes glued to the habitat as I powered south, I failed to appreciate thunderheads gathering to the north. The skies darkened as I approached Coalmont, and I was caught in another lightning storm before I realized it. I spotted a house a half-mile down the road and hustled toward it. There weren't any cars in the driveway, so I assumed the homeowners were elsewhere. Figuring I'd shelter and depart before they returned, I leaned my bike against the house and turned toward the adjacent barn, where I planned to ride out the weather.

As I did that, a dog went berserk inside the home. I saw a human form stirring, and I hoped I wasn't about to come face-to-face with a baseball bat, branding iron, or firearm. I'd found forthrightness the best recourse, so I readied my rap as the shadowy form approached the door. The portal opened, and a middle-aged woman stuck her head out. She peered at me through a set of thick, dated glasses.

"Hi! I was just biking through when the storm popped up," I said. "I was going to hunker down in your barn, but I'll get going. Sorry to have bothered you."

She squinted to have a better look at me and spoke confidently. "Don't be ridiculous. Put your bike in the barn and come into the house."

Encouraged by her demeanor and invitation, I complied. She introduced herself as Sophie, instructed me to sit on the sofa, and fetched some lemonade from the kitchen. She was curious how I'd wandered onto her isolated doorstep, so I explained my bird-motivated travels in between swigs of tart nectar.

"What kind of birds are you looking for here in Coalmont?" she asked.

Avoiding unnecessary specifics, I replied, "Sage-grouse. They live all through here, but they're sneaky and hard to find."

Sophie's eyes opened wide. She replied, "I know those birds! There's a lek right down the street. They aren't on it now, but I occasionally flush individual birds when I'm riding my horse." She continued, "We'll let the storm pass and go searching together. I'll make you lunch in the meantime. Is pasta OK?"

My promotion from trespasser to lunch guest was sudden, but her hospitality was more than welcome against the thunderous alternative. Moving to the kitchen, our exchange revealed Sophie was an Iowa native who enjoyed cycling in her youth.

"I'd often get caught in surprise storms. Barns were the perfect shelters, so that's where I went," she said. "I recognized your predicament immediately. Not too many murderers arrive by bike, so I figured the percentages were in my favor."

When I asked how she landed in Coalmont, a twenty-person town that made Iowa look like the most bustling place on Earth, she explained she'd followed a man to Denver. Yearning to escape those urban confines after the relationship cratered, she bounced around Colorado before meeting her current husband, Brian, and settling in Coalmont. She beamed as she described how the two of them designed and built the gorgeous log house in which we were sitting.

The rain abated after lunch, and Sophie and I explored her breathtaking acreage under blue skies. She was on horseback, I was on foot, and her long-legged mutt bounded along in between us. We pushed through several miles of suitable habitat, but we didn't intersect any sage-grouse, a result reinforcing my growing understanding of the species' historic decline and summertime dispersal.

When I referenced departing, the remaining forty-five miles to Steamboat Springs looming larger as the afternoon waned, Sophie insisted that I spend the night and resume the search on a different piece of property the following morning. At that moment, finding the bird was more important than making miles, so I decided to overnight rather than press forward. I was curious to meet Brian after spending the afternoon with Sophie, so there was personal interest as well.

After Brian arrived home, dinner was fajitas, and our accompanying conversation wound through politics, education, and travel. Sophie and Brian extolled their Colorado surroundings and reiterated how nice it was to live in isolation, their nearest neighbor a half mile away.

"Outside of work, we're basically hermits," Sophie said.

I found that label incongruous given their gift for gab and capacity for compassion. I asked about their decision to forgo cable television and the internet, a unique choice among my hosts to that point.

Brian answered, "Who needs a hundred channels of trash? Books are so much better. Besides, the Broncos and Rockies are on local TV if I want to watch a game."

Sophie added, "And the internet is mostly nonsense. I prefer the newspaper, both in content and feel."

Insulated from cable news, social media, and commercial bombardment, their perspective seemed authentic and uncorrupted. They were Colorado's best kept secret, pure as the mountain air they breathed and refreshing as the well water they drank. Much like the sage-grouse that graced their property, the reclusive couple could dazzle when encountered under appropriate circumstances.

Lying in bed after dinner, I hoped I'd find contentment equivalent to theirs. They didn't need recognition or validation, and they'd oriented their lives around internal definitions of happiness rather than external suggestions of status or success. Sophie and Brian were at least twenty years my senior, so I figured age influenced their perspective. Geography probably contributed, too, Coalmont imposing less expectation than Palo Alto, New York, or Boston, but I suspected their bond to have additionally shaped their attitudes. At ease together, they focused on each other and the activities that brought them shared joy: riding horses, exploring the outdoors, and reading books. They'd avoided consumer- and misinformation-cultures, and they were an inspiring picture of happiness as a result. Thankful that the lightning storm had illuminated them for me, I drifted off.

Despite four hours of searching, Sophie and I were unable to scare up a sage-grouse the following morning. She invited me to extend the search into the afternoon and spend another night, but I declined. I needed to keep moving ahead of a commitment to meet another birder in Craig, a day beyond Steamboat Springs. As I left, Sophie offered me her phone number in case I ever passed through the area again.

"Thanks a lot, Sophie. And what's your last name? I need it for the contact list in my cell phone," I said.

She answered without hesitation. "Put me in as Sophie Sage-Grouse. That way you'll never confuse me with someone else."

There was zero chance of that happening. Hermit, saint, or genius, she was Colorado gold, an independent spirit and unique wit concealed in an endless sea of sage. My inability to find the sought bird before encountering her was a blessing, and I bid her goodbye before rolling down the driveway. I rejoined Highway 14 south, climbed out of the basin, merged onto Highway 40 west, and dropped into Steamboat Springs for the night.

I spent two nights in the mining town of Craig, the intervening day used to add Sharp-tailed Grouse, and continued ninety sweltering miles along Highway 40 to reach Dinosaur, Colorado, on July 18th. Named for the abundance of skeletons unearthed in the vicinity, the tiny town boasts dinosaur murals and statues and features streets like Brontosaurus Boulevard and Triceratops Terrace.

Correspondingly prehistoric, my decaying motel offered access to several alleged strongholds for my newest nemesis, the Greater Sage-Grouse. Utah's Diamond Mountain looked promising, but the ride to the elevated mesa would be a colossal chore; the second half of the 50-mile slog would require 3,800 feet of climbing in ninety-degree heat.

I left Dinosaur at sunrise, wheeled into Utah for the second time, and crossed the Green River at Jensen. From there, I departed Highway 40 and began my climb on crumbling county roads, my pace slowing as the course steepened. The region was as dry as any I'd visited, and the rocky landscape radiated heat as the sun intensified. By the time I reached the final, five-mile pitch to Diamond Mountain, I was pouring sweat. Reeling and light-headed, I wondered what future paleontologist would unearth my fossilized remains—feet still clipped to pedals—from the dusty surrounds.

I outlasted searing heat and crippling exhaustion and reached the rim of the plateau at noon. Gazing south off the precipice, I felt like I was standing on the edge of the world, heat waves blending roads and topography into a swirling milieu for as far as I could see. Reorienting north, away from that confusion, I faced a vast swath of sagebrush backed by craggy peaks. I knew there were sage-grouse in my field of view, but I had to scour enough habitat to find them.

Diamond Mountain's sagebrush was the thickest and woodiest I'd explored. That was an encouraging sign given that sage-grouse prefer old-growth plants, so navigating the habitat was a prickly process. Bloody scrapes on my legs indicated progress, but I'd hardly moved in the plateau's immense context. The afternoon sun beat down with vengeance, and the heat sapped energy with each plodding step. No path was better than any other; randomly schlepping through the landscape while clapping and shouting was my only recourse. Drained of motivation, patience, and water after ninety minutes, I made a wide turn to double back to my bicycle.

Halfway through my unenthusiastic return, two Greater Sage-Grouse exploded from the brush in front of me. A celebratory handclap flushed four more, and a jubilant cackle revealed a final five, the last photographed before it disappeared into the distance. Arms extended overhead, binoculars dangling from my neck, and camera swinging from my shoulder, I sounded a triumphant yawp. There was no way Sophie could have heard me, but I did my best to believe she did.

My week-long quest had culminated in one of the year's most savored victories, and I was thankful that my self-powered course had prescribed such effort. While I would have loved the opportunity to witness the springtime courtship spectacle, my protracted pursuit reminded me that the easiest route is rarely the most rewarding. The bicycle was an immeasurable pain in the ass at many points, but I understood that my ability to push through challenges would impart my undertaking with lasting value. The week's mission accomplished, I descended to Vernal, Utah, for the night, the 200th of my self-powered journey.

NINETEEN

The Dating Game

Endorheic lakes form in basins lacking natural drainage; without streams or rivers for outflow, water escapes such bodies only through evaporation, a protracted process that concentrates ionic salts dissolved in the original inflow. If continued over millennia, the concentrated minerals precipitate as solids, a phenomenon exemplified by Utah's Great Salt Lake. Formed 11,000 years ago, the contemporary lake looks like the world's largest serving of shaved ice. Gazing at the crystalline slurry, I struggled to reconcile that image with the suffocating heat as I crossed the low causeway to Antelope Island on July 25th.

Birdlife was plentiful in the salty surroundings. Most abundant were Red-necked Phalaropes, the dainty birds paddling around and plucking flies from the surface of the lake. Bigger than sparrows but smaller than robins, phalaropes are specialized sandpipers that prefer swimming to walking. One of just three phalarope species, the grayish Red-necked nests on Arctic tundra and winters in tropical oceans. Despite their long commutes to and from the Arctic, their time at high latitudes is brief; the birds had finished nesting and were already returning south at our late-July intersection, the Great Salt Lake a bounteous refueling stop on their equatorial return.

Though their aquatic affinities are curious, phalaropes are most notable because they reverse the stereotypical gender roles; females are more brightly colored and compete for male mates. Once a female lays eggs, she departs and mates with additional males in a sequentially polyandrous manner; each abandoned male incubates the eggs and raises the brood in her absence. In this way, and in complete opposition to the polygynous Greater Sage-Grouse, female phalaropes drive the population's genetics. With tens of thousands of Red-necked Phalaropes in my Great Salt Lake view, their atypical mating system seemed to function as well as any other.

Departing Antelope Island, I continued thirty miles to South Ogden, where I met up with a cycling couple I met online. Ammon, the older of the pair, was tall and fit; with graying hair and emerging wrinkles, he appeared in his late forties. Thomas was slightly shorter, a boyish visage suggesting he was in his mid-thirties. They prepared an elaborate dinner at their home, and we enjoyed lively conversation on their back patio.

During dessert, I gleaned that Ammon had a twenty-something daughter. He and Thomas had been open about their relationship, so I asked how she fit into the picture. Ammon swigged his glass of wine, refilled it, and explained. "I grew up in a hardcore Mormon family, and being gay isn't part of the program," he said.

"I can't imagine how difficult that was for you," I replied.

He continued, "Well, when I was eighteen, I went to my priest, revealed that I was attracted to men, and asked for advice. He told me to keep my feelings secret and said they'd go away if I found a wife."

Ammon had trusted the church at other confused points, so he found a woman, Maylyn, with whom he was compatible and got married.

"And let me guess, your feelings didn't change?" I asked.

"Hell no," he replied. "But the next step in the prescribed cure was kids. So we had three. I love my family, but the whole thing was so confusing. It took a while before I could tell them the truth."

"How long before you came out?" I asked.

"Seventeen years."

My heart sank. He'd endured nearly two decades of emotional conflict at the hands of institutional dogma and intolerance. Maylyn was shocked to learn Ammon was gay; she, like everyone outside the marriage, believed he was heterosexual. Despite her own pain, she was understanding, even supportive, as they divorced. Ammon subsequently left the church and began to live his life as an openly gay man. Ammon and Maylyn remained close, and, in a heartwarming twist, several years post-separation she officiated the commitment ceremony between Ammon and Thomas.

Turning to face Thomas, Ammon added, "It was a long road, but it was worth it."

Thomas reached over and put his hand on Ammon's forearm. Seeing Ammon with Thomas, seeing the love in how they interacted, I knew their paths had converged at the best possible time. I thanked Ammon for sharing his story with me and retired for the night, the day's eighty-four superheated miles rendering me asleep as soon as my head hit the pillow.

I bid the pair goodbye early the following morning and continued north toward Logan. Thinking about Ammon's story as I climbed up Ogden Canyon and powered around Pineview Reservoir, I recalled the other couples I'd met along my seven-month arc: Bobby and Susan in Connecticut, Ralph and Mary in Alabama, Ron and Janet in Arizona, and Sophie and Brian in Colorado. I was too late to meet Bernie's wife while I was in Florida, but the memories he shared suggested love perdures in ways that bodies do not. Each of those people had found their match in another, and I realized human attraction is the most interesting on Earth. Intriguing as are the ways of grouse and phalaropes, no courtship is as individualized, protracted, or entertaining as human pairing. Finding one's reflection in another person is infinitely more challenging and rewarding than obeying a biological breeding cycle.

———————

Among dozens of bars within stumbling distance of the East Village apartment I occupied while attending NYU, Professor Thom's on 2nd Avenue and 13th Street emerged as my favorite because of its unabashed loyalty to the Boston Red Sox, a team for which I'd cultivated an unlikely affinity. Growing up in Philadelphia, I'd formed deep bonds with my hometown teams—Phillies, Flyers, 76ers, Eagles—while establishing parallel disdain for their New York counterparts, none more despised than the Yankee corporate juggernaut that purchased four World Series titles between 1996 and 2000. My underdog affinities made me wary of such power consolidation, so I adopted the Yankee's main rival, the underachieving but charismatic Red Sox, alongside my childhood Phillies when I moved to Boston in 2001. That allegiance accompanied me to New York when I matriculated at NYU in the fall of 2004.

Professor Thom's was therefore a friendly Red Sox port in an otherwise enemy Yankee sea, and I became as much a fixture at the Boston-themed establishment as pints of Samuel Adams or photos of Larry Bird. I befriended many regulars, and the bartenders grossly undercharged me because they knew I was a broke graduate student. With television and air-conditioning, two amenities my threadbare apartment lacked, the bar was my de facto living room. The watering hole was only 200 yards from my 14th Street hovel, so stumbling home was easy.

I didn't expect the night of Wednesday, October 29th, 2008, would be any different save for one consideration. I wasn't at Professor Thom's to watch my adopted Red Sox because they'd been eliminated from the playoffs. I was there to watch my childhood-affiliate Philadelphia Phillies face the Tampa Bay Rays in Game 5 of the World Series, a best-of-seven affair in which the Phillies led three games to one.

More specifically, I was at the bar to watch the resumption of Game 5. The contest had started two nights earlier, on October 27th, but torrential rain in Philadelphia had forced the stoppage of play midway through the sixth inning with the score tied at 2–2. That pause represented the first time in World Series history—607 individual games spread across 105 years—that a game was suspended without resolution. The deluge continued through the 28th, and play was scheduled to resume when the rain stopped on the afternoon of the 29th. If that historic rainstorm didn't delay that specific baseball game, my life would have assumed an entirely different course.

Ten minutes after I pulled up to the bar to watch the remainder of Game 5, a dark-haired woman claimed the barstool two removed from my own. She was alone, didn't display a wedding ring, and wasn't obsessively checking her cell phone as though waiting for someone. Fortified by the five beers I'd consumed at my apartment as a pregame warm-up, I leaned over the empty seat between us.

"Just to warn you—I'm a die-hard Phillies fan," I said. "If you're a Rays fan, then we're going to have to flip a coin to see who relocates."

She tossed her hair over her shoulder, turned toward me, and spoke through a luminous smile. "No worries. I'm a Red Sox fan since I went to college outside Boston. I'm not invested in this game. I just like watching the World Series."

I nearly fell off my stool; in one sentence, she'd morphed from bar-stool bombshell to friendly and educated woman with Red Sox affinity to boot. She was too good to be true, but I kept my cool and tried not to appear too desperate.

"Nice," I replied. "The Sox are my other team since I lived in Boston before I moved here. I've always hated the Yankees, so it makes sense, right? You must have been here before if you're a Boston fan?"

"Yeah, I live right around the corner," she explained. "My roommates let me have the TV for the other games but not for this rescheduled fragment since they're huge *Lost* fans. This is the best baseball bar around, so I came here to watch."

Given how much time I spent at the bar, I was shocked I hadn't run into her before. It was possible I'd met her in an inebriated stupor, but her willingness to engage me suggested I was working with a clean slate.

"What's your name?" I inquired. The game's play-by-play piped through the bar's stereo at that instant.

She replied, but I didn't hear her against the sudden injection of noise. I made my best guess, "Silvia?

She shouted, "Sonia! S-o-n-i-a!"

Smiling, she returned the question.

I answered with gusto, "Dorian. D-o-r-i-a-n! Can I buy you a beer? B-e-e-r?"

She laughed again. "Sure, why not?" she said with a smile.

The drama of the resumed game grew, and my drinking pace and conversation with Sonia gathered corresponding steam. Her lips were voluminous, and she flashed a pearly grin whenever one of my jokes landed. I had no idea if she was laughing with me or at me; she was still sitting there, and that's all that mattered to a drunk like me.

Unfortunately, another guy, Ray, had weaseled his way into the empty seat between Sonia and me. He was nice enough, but he functioned as distraction and obstacle as I tried to engage Sonia because he kept inserting himself into the conversation at awkward junctures. Seizing the opportunity when Sonia excused herself to use the restroom, I leaned over to Ray with a proposition.

"How about I buy you a beer and a shot, and you move to the other end of the bar? Offer disappears the instant she gets back."

A deal struck, Ray tottered away in exchange for a bottle of Bud Light and a shooter of Jim Beam. Upon returning, Sonia asked what happened to Ray.

Pointing toward our discarded third wheel at the far end of the bar, I said, "Two of his friends rolled in, so he joined them."

"Oh well, guess it's just us!" she said.

We picked up right where we left off, Sonia laughing at my stupid jokes and me wading deeper into her brown eyes. Her beauty was exceeded only by her engaging personality, and the two-day rain delay felt more fortuitous with each inning.

The Phillies held on for a 4–3 win and thereby captured the 2008 World Series. I put away a rapid succession of celebratory shots before stepping out front for a victory cigarette. I hadn't explicitly expected Sonia to be inside when I returned, but I was surprised when she strolled out while I was savoring my smoke. She congratulated me on the Phillies's win and said she was heading home because she had a big day at work tomorrow. Mentally compromised under the combined influence of Budweiser, Goldschläger, and menthol cigarettes, I bid her a cordial but routine goodnight.

Watching her make her way down the street, my brain overcame those dulling influences.

What the hell am I doing? She's perfect! I can't let her get away without getting her number!

My window of opportunity was closing, so I ditched the remainder of my cigarette and stumbled down the block. Traffic stalled her retreat, and I caught up to her while she was waiting at the corner, the red glow from an overhead Kentucky Fried Chicken sign imparting her with unanticipated radiance. Through slurred speech and glassy eyes, I made my best pitch.

"Hey, I had a really good time with you. Any chance I could call you sometime?" I asked.

"Sure, I had fun. We should do it again," she said.

I expected I'd need to deploy additional wit and charm to obtain the desired digits, so her expedited offer had me searching for my phone and scrambling to open my contact list before she changed her mind. I was still fumbling around when she began dictating her number. I didn't recognize the area code and interrupted her midstream. "Wait a minute," I asked tentatively. "Area code 562? Is this a fake phone number?"

"No, it's an LA number," she said without inflection.

I was relieved; she hadn't seen through my embarrassing reveal. She reinitiated her dictation but stopped suddenly. "Hold on. Did someone seriously give you a fake phone number?" she asked.

I had shown too much of my hand. I struggled to regain my mojo. "No. Well, errr, maybe," I said. "Anyway, what's that number again?"

"Sounds like you have some good dating stories! My dating experience has been a train wreck, but we'll save those details for next time. This is my real number, I promise," she said with a laugh.

I entered it into my phone and told her I'd call her next week, after I returned from the Phillies's victory parade. Future communication secured, we gave each other a quick hug. She disappeared around the corner, and I returned to the bar where I drank myself into my usual oblivion.

I went to Philadelphia for the weekend but spent more time thinking about Sonia than the Phillies. She was a splinter in my mind, and a volley of playful texts connected us while I attended the championship celebration, a bottle of vodka tucked into my coat as I roamed the city streets on that jubilant occasion.

I returned to New York on Sunday night and made plans to meet Sonia at Professor Thom's on Monday, the eve of Barack Obama's election. I departed the lab early evening, grabbed two slices of pizza on my walk home, pounded a bunch of beers while playing records at my apartment, and headed out to be at Professor Thom's at the prearranged eight p.m.

Sonia entered the bar and was even more radiant than I remembered from our first encounter. We waded through with pleasantries, and within minutes I was adrift on her every word. The conversation was fluid as we dove into each other.

"What do you do again? You're a scientist or something?" she asked with an incredulous look.

I replied, "Yeah, I'm a PhD student at NYU. I study cell polarity and embryonic morphogenesis."

"No way! I thought you were joking the other night!" she said.

"Yeah, it's totally legit. I'm hoping to end up as a college professor, but it's going to take a while."

"You don't seem like a scientist," she said. "You're kinda goofy, like a clown or a comedian."

I lived in perpetual fear of being taken too seriously, so I received those comparisons as compliments. I doubled down on my appointed role. "Well, you seem to like laughing, so I'll stick to what's working for now," I said. "Gotta keep the material fresh so you don't split!"

She laughed and subsequently chronicled her own life trajectory. Raised as one of four Mexican-American siblings in East Los Angeles, Sonia attended Gordon College north of Boston and double-majored in communications and visual arts. Her biracial upbringing motivated a minor in Spanish, and her year abroad in Spain reinforced her respect for multiculturalism while seeding a passion for international travel.

After graduation, Sonia interned as a set builder in a community theater, waitressed at an Italian restaurant, and worked as flight attendant for Eos, a business-class carrier that made daily hops between New York and London. Laid-off when the Great Recession bankrupted the airline earlier that year, she rebounded at a corporate travel agency, where she managed the hotel program for Condé Nast Publications.

"My job is pretty rough," she explained. "I travel around the US and Europe and decide which hotels Condé should use for its high-end and celebrity clientele."

Beyond her baseball leanings, I discovered Sonia was a rabid ice hockey fan, another plus given my strong affinity for the sport. She also enjoyed hiking and camping, activities that would dovetail nicely with my birding interest, if I ever resurrected it. She wielded a spellbinding balance of confidence and vulnerability, and her blend of sincerity and sarcasm was magnetic.

By the conclusion of our three-hour conversation I was totally smitten, and Sonia's undivided attention suggested the attraction was mutual. Leaning in for a kiss, I was well received, her lips as tender as I'd imagined for the previous five days.

We parted ways after that intimate moment, Sonia toward her apartment and I toward another bar and my inevitable blackout. She was the most beautiful phalarope in the world, and I desperately wanted to be the main man in her life. Falling as hard for her as I had for alcohol twelve years earlier, I hoped she felt the same.

TWENTY

Carpe Diem

My reminiscence was interrupted by gunfire just outside of Liberty, Utah. Continuing cautiously ahead, I approached a shooting range at the base of a steep ridge. While I paused to rest and rehydrate ahead of my anticipated ascent, a pickup truck turned out of the range parking lot. The driver's window opened, and a forty-something man in a cowboy hat engaged me.

"Where ya headed?" he asked with a smile and Western drawl.

I pointed toward the hill and replied, "Logan. I think this is the way, but I thought the road would be paved."

"Yep, that's the way. It's dirt and a hell of a climb. Happy to give you a lift. Won't take ten minutes in this beast," he said as he tapped the outside of his truck with his left hand.

"Thanks, man. I appreciate the offer, but I gotta do this on my own. I haven't taken a ride across seven months and 9,000 miles," I said proudly.

The passenger leaned over and spoke, "Hot damn! You gotta have the strongest legs and sorest ass on Earth! All that time, just you and the bike? You carryin'?"

I replied, "Nope, but I have a small utility knife."

Both of their faces froze, but the driver first regained the capacity for speech. "Too many crazies out there to go anywhere without a gun," he said. "You're a sittin' duck. Be careful."

"Will do," I replied.

The pair drove off, and I began the steep climb. The road was rocky and rutted, and my tires, designed for pavement, slipped on loose earth. Unable to maintain momentum, I dismounted and pushed. The track resembled a backcountry hiking trail toward the top, and I leaned into the hillside to prevent my bicycle from rolling backward. My back aching and my legs wobbling, I was relieved to reach the crest, where a graded track invited a smooth and gradual descent.

Rolling downhill, I revisited my exchange with the pair in the pickup. I had, while planning my adventure, considered carrying a small firearm; it wouldn't have added much weight, and I imagined situations where it could be the difference between life and death.

However, with deeper consideration, I realized a gun wouldn't mitigate the overarching vehicular danger I'd face. I also thought carrying a gun would steep my journey in fear, a reactive position from which bad decisions are made. I didn't want to take a life because I was scared, possibly for unfounded reasons. Security is a reciprocal function, so I decided I would treat people as friends instead of enemies. I hadn't questioned my decision to be firearm-free by the time I reached Logan, and I was confident I'd be run over long before I was forced to reach for a gun I didn't have.

Continuing northeast on Highway 89 the next morning, July 27th, I crossed the Bear River Range before dropping to Bear Lake and continuing into Idaho. My brush with the Gem State was brief, and I continued into Wyoming the following day. Powering toward Grand Teton National Park, I hoped to score Trumpeter Swan, Ruffed Grouse, Great Gray Owl, and Black-backed Woodpecker, boreal birds that reach into the region from their mostly Canadian ranges. The swan showed in a roadside pond as I powered into Jackson on July 29th, so I'd have four days to find the other three, the owl particularly prized.

Thirty inches long and sporting a five-foot wingspan, the Great Gray is the world's most imposing owl despite weighing less than three pounds, insulating feathers accounting for most of their perceived bulk. Their beady yellow eyes demand attention, but their exaggerated facial discs are their most characteristic feature. Those evolved structures funnel sound waves toward the bird's ears and allow a Great Gray to hear a scampering mouse from a hundred yards.

Great Grays are largely diurnal, so I used the afternoon of my arrival and my entire second day to pursue the bird. Though I observed porcupine, fox, elk, moose, bear, and Black-backed Woodpecker while I wheeled through woodlands and meadows, the owl proved elusive. Enlisting the assistance of a local biologist on my third day, we flushed a candidate as we pushed through a grove of pines. The bulky form flew across a small clearing, and we crept closer to confirm our Great Gray suspicions. Scanning the trees, I seized on the trademark facial discs, a pair of citrine eyes boring through my binoculars and into my soul. My camera immortalized the arboreal noble before he fled, but a digital representation would never capture his grace and majesty. Grateful for my glimpse of the powerful predator, I retreated to an ice cream parlor and preyed on two scoops of chocolate–peanut butter.

Unable to find Ruffed Grouse after investing an additional day around Jackson, I pushed west out of town on August 2nd. I followed Highway 22 across the Snake River and gained elevation toward Teton Pass. The final four miles of that climb, an insane stretch during which I overcame 2,000 vertical feet, was the most challenging pitch I'd experienced, Colorado's Guanella Pass included. My throbbing quadriceps felt like they were going to tear like paper, and I feared my heart might explode as the overloaded organ struggled to meet my metabolic demands.

The test abated 8,431 feet above sea level. I dismounted, removed my sweat-soaked shirt, and collapsed onto a boulder. Feeling as ambitious as the lichens that were decorating my temporary support, I surrendered to a pair of overhead Turkey Vultures. Their interest only superficial, I lolled in half-sleep for the next twenty minutes. Eventually able, I stood and gazed west into Idaho, my home for the following week.

Zipping downhill, I arrived at Teton Spring Farm an hour beyond Victor. Sarah was unmistakable, her wavy, straw-colored hair blowing in the Idaho wind as I approached the farmhouse. High school classmates at Hotchkiss, Sarah and I last crossed paths in New York, when I was at NYU and she was at Parsons School of Design. We were friendly when we bumped into each other at alumni events but didn't communicate outside of those orchestrated encounters, so I wasn't aware of her cross-country move until a mutual friend informed me. Her metamorphosis from urbanite to farmer intrigued me, and I asked about her evolution as she toured me around the biodynamic dairy that she ran with her husband, Frank.

Sarah spoke with a scratchy but melodious warble. "My sister moved to Sun Valley, Idaho, in 2005," she said. "I visited her that fall and fell in love with Idaho's landscapes. I knew right then I wanted to live here."

She returned to New York, packed up her stuff, sublet her apartment, and moved in a month. Her motivation was admirable, but I didn't understand her rush. Idaho wasn't some trendy opportunity that would disappear if not immediately seized, so I asked why she'd abandoned her career course and social network so abruptly.

"Because I could have talked myself out of it with additional time," she said. "I felt this personal manifest destiny to go west and start a new life. I wanted to use the enthusiasm before it dissipated. 'Carpe diem,' right?"

Sarah recounted how, upon relocating to Idaho, she worked for the Sun Valley–Ketchum Chamber of Commerce before the Great Recession killed the position. She bounced around after that downturn but eventually landed in Jackson, where she worked as the communications director for an independently owned and environmentally conscious grocery store.

She beamed as she continued, "Frank sold his milk at our store. We met, one thing led to another, and before I knew it, I was moving over Teton Pass to help him run the dairy."

Sarah shepherded me into a pasture where a dozen Brown Swiss cows were grazing. Friendly to the touch, each had a name, a recognizable coat pattern, and a distinct personality. They were members of the family, and from them flowed the milk that kept the farm afloat. With a carpet of grass beneath their hooves and mountain air circulating through their lungs, their existence was a refreshing contrast to the industrial feedlots I'd encountered at other points in my travels. In those, thousands of animals languished in a putrid mix of mud and feces, the stench gut-wrenching from a mile away.

Frank emerged from the milking facility abutting the pasture. He wasn't the burly, corn-fed giant I expected; his short stature, wiry frame, shoulder-length blonde hair, and blue eyes suggested him to be a California surfer before an Idaho farmer. My grasp of his extended hand dispelled that notion; grease stains and broken nails revealed his vocation was anything but a day at the beach. He invited me to make myself at home before ducking back into the facility. More a man of action than words, I could see why Idaho Sarah had fallen for him.

I spent the following morning, August 3rd, recovering from my Teton trial but rallied for some afternoon birding. Pushing through a stand of alders behind the barn, I flushed two Ruffed Grouse, negating my multiple misses in Jackson.

While I birded, I kept a curious eye on Frank. He'd started work before sunrise and was in constant motion between the pasture, milking facility, and barn. As Sarah and I sat down to dinner, Frank fired up his tractor and began baling hay, a task he'd complete overnight due to approaching rain. He was still working when I flicked off my light at eleven p.m., and I realized Frank's notion of normal business hours was as warped as my understanding of an agrarian existence.

Raised as a child of privilege and groomed for professional pursuits from an early age, I never gave farming and ranching much thought. I took the production of fruit, vegetables, milk, and meat for granted because I was removed from the providers, and I disregarded agrarian endeavors as defaults for people without intellectual ability or drive. Stereotypically elitist, my perspective was rooted entirely in circumstantial ignorance; had I experienced the other side of the urban–rural divide, I suspect my views of farming and ranching would have been different.

My time at Teton Spring was an overdue reality check. Sarah was born into similar privilege but departed it to become a farmer. I wasn't familiar with Frank's background—he was too busy to socialize the entire time I was there—but his work ethic suggested he'd be successful in any arena. Each could have done something else, but they chose to build a dairy together, a herculean task that required a labor and a love equal to anything I'd invested in my scientific career. Sarah and Frank forced me to recalibrate my perspective and assign appropriate value to the contribution ranchers and agrarians make to society. Though I'd done so in the past, I wouldn't again underestimate the spirit or ability of the American family farmer.

I thanked Sarah for the window into her world and departed the farm as the forecasted rain materialized. Frank had worked through the night but paused to wish me well as I exited the property and joined Highway 31. Steady precipitation continued through midday but was redeemed by an east tailwind in the afternoon. Those facilitating gales sped me along Highway 26 and animated oceans of roadside wheat. Adrift on amber waves of grain, I surfed west.

I reached Idaho Falls late in the afternoon, ditched my panniers at my motel, and initiated an evening search for Gray Partridge in the agricultural swath west of town. It took the better part of two hours—and an additional twenty-two miles of riding beyond the seventy-two I covered to Idaho Falls—but I flushed a covey of ten birds from the road margin as sunset approached.

More than being species #514, the partridge was the last bird I needed to find during my mountain summer. Since joining the Rockies in northern Arizona in early June, I'd found every species for which I'd searched, plus a few I'd overlooked or thought impossible. Extending my memory, I realized I'd found every sought species since departing the Texas coast in late April. Given that perfect run, my decision to fold the final three trans-Gulf migrants seemed forgiven; had I delayed even one additional day, I would not have intersected everything I subsequently did, at least not on the same expedited timetable. Anxious to extend to my birding fortune, I looked forward to the Pacific Coast.

That destination was 800 riding miles from Idaho Falls, and I covered 313 of those to Boise across the next four days. Made mostly on Interstates 84 and 86, those miles put me at over 10,000 for the year and gave me a wonderful sense of southern Idaho. While not as spectacular as Arizona or Colorado, the rocky hills and golden grasslands were stunning in a subtler way. There was little development beyond the interstate, and I felt trapped in time as I gazed at the unspoiled landscape with my right eye while watching for approaching trucks with my left. Caught between centuries, my bicycle seemed an appropriate vantage to appreciate the contrast.

I took a rest day in Boise and used the downtime to consider my course for the next few weeks. My imagined path would have routed me across Oregon and onto the Pacific Coast by early September, but I'd rebuilt my time buffer to two weeks by that juncture, August 9th. My legs felt strong after surviving the Rockies, so I contemplated cashing in some of my accrued time to search for bonus birds.

Cross-referencing road maps with range maps, I discovered I could add Spruce Grouse and Boreal Chickadee, two northern species I'd discounted in the planning phase, if I extended my route to Washington's North Cascades. Subsequent time on Puget Sound would guarantee Mew Gull, a species I'd struggle to find farther south, and a stop at Mount Rainier would afford opportunity for Gray-crowned Rosy-Finch, a relative of the Brown-capped variety I'd observed on Mount Bierstadt in Colorado. If I followed the Washington coast south after departing Rainier, crossed the Columbia River, and rejoined my original route on the Oregon coast, then I'd cycle at least 700 unanticipated miles in pursuit of the four extra birds. The extension would be a huge undertaking, but at 175 miles per species, it would be less than the dedicated 300 I sank into Greater Prairie-Chicken in eastern Colorado.

Irrespective of my ability to find the four target birds, the route extension would deplete my time buffer and leave me no cushion against future sickness, injury, or fatigue. My Rocky Mountain triumphs had paved a viable path to 600 species if I did nothing more than complete my imagined route, so any deviation from that trajectory would jeopardize my run at 600 more than facilitate it. If finding 600 species was my ultimate goal, then I should have discounted Washington and proceeded west across Oregon as planned.

But it wasn't. Although the birds gave the journey purpose and shape, they were a convenient scaffold on which I'd hung a more elaborate and meaningful search for self. At my journey's core was facing uncertainty and overcoming challenges, and I knew I needed to grow the project beyond my original vision to maximize its impact. Worse than not completing my route would be leaving opportunity on the table. I'd never be in the same position again, and I understood the conservative path—the same path that had threatened to hold me prisoner in my lab—was where dreams died. Heeding Sarah's inspiring call to action, I steered toward Washington before I could talk myself out of it.

TWENTY-ONE

Agency

Rocky outcroppings interrupted rolling, grassy hills, and heat sucked moisture from my pores as I struggled along the shoulder of Interstate 84 on August 11th, two days after departing Boise. I didn't expect much human presence in northeastern Oregon, a necessary hurdle as I aspired to Washington, but the degree of isolation shocked me. My course recalled the nothingness of West Texas, and I counted mileposts as I battled the topography toward Le Grande, 120 miles from my Ontario starting point.

Halfway through that ten-hour slog, a Red-tailed Hawk took off from a roadside fencepost. The barrel-chested bird took several powerful flaps, caught a midday thermal—a rising column of hot air that functions as an invisible escalator—and spiraled skyward. Envious of the bird's effortless ascent while I labored through the parched landscape, I recalled a previous intersection with the species.

My commitment to alcohol and drugs relegated birding to an infrequent distraction through my first four years of graduate school, occasional outings with my advisor my only avian engagement, but I introduced Sonia to my former passion a month into our association. She'd spoken longingly of hiking and camping on several occasions, and I thought a birding outing would be a nice change from the bars and house parties we'd frequented since our World Series introduction. My suggestion of a walk through the Ramble, a wooded area of Central Park known for sexual shenanigans, was met with playful apprehension.

"The Ramble? Really? What kind of woman do you think I am?" Sonia asked in jest.

I replied, "I dunno, but I guess we'll find out when we get there!"

She laughed, "I just can't say 'I'm going birdwatching in the Ramble' with a straight face!"

My exaggerated cajoling overcame her dramatized resistance, and we formulated a plan for the upcoming weekend. Among the birds I hoped to show Sonia was the Red-tailed Hawk. Chocolate-brown above and cream-white below, the species is named for the rufous tones on the topside of the tail. The bird is common throughout North America and—unlike the specialist Greater Sage-Grouse—has displayed remarkable adaptability in the face of human encroachment. The consummate generalists, Red-tails have taken to farms, suburban neighborhoods, and inner cities. Central Park's inexhaustible ranks of pigeons, squirrels, and rats have historically supported several resident pairs, and Sonia and I headed for a known Red-tail area after exiting the subway.

We hadn't been in the park for five minutes before I spotted an immature Red-tail perched in a tree. The handsome raptor was peering intently at something below it. As though cued for Sonia's amazement, the predator suddenly descended on an unawares pigeon and sank its talons into the prey's back. The pigeon flapped frantically, but the hawk held tight and increased its advantage when it wrapped one foot around the pigeon's head. We crept closer as the helpless prey succumbed, Sonia gripping my right arm for reassurance.

"Oh my god! It's ripping the poor thing to shreds!"

"I know. It's great, right? Just like *Animal Planet*," I replied.

Advancing to within twenty feet, we watched the Red-tail eviscerate its breakfast, the blood of the disemboweled pigeon seeping onto the pavement. I couldn't have imagined a more enthralling birding introduction, especially with Sonia holding onto me for comfort.

Extending our exploration after the hawk flew off, we ambled into the Ramble and navigated a network of paths en route to a central collection of bird feeders. We snuggled on a bench, and I gave Sonia my binoculars and held her hands as she familiarized herself with the focusing wheel. That mechanism mastered, I pointed out the different birds for her.

"That gray one with the black cap is a chickadee, and the yellow one with black wings is a goldfinch," I said.

"Chickadee, check. Goldfinch, got it," Sonia said. "Ooooh! What's the one over there, with the red?"

"Good eye," I answered. "That's a Red-bellied Woodpecker. They love those dangling suet blocks."

Sonia was captivated by the high-wire woodpecker. No matter what foothold he employed, the suspended buffet twisted and swayed as he pecked at it. His trapeze act had us both smiling, Sonia at the bird and me at her.

Continuing to the Lake, we watched a group of Northern Shovelers turn their characteristic foraging circles. Named for their splayed bills, the ducks feed by swimming in synchronized loops, those rotations stirring up muck from which they filter plant and invertebrate edibles. Males are white with green heads and chestnut flanks, females are a beautiful assemblage of browns, and the twenty pairs created a mesmerizing cycle as they orbited one another, their swirling formation like a kaleidoscopic hurricane as they swept under our Bow Bridge vantage.

"This is wonderful," Sonia said. "I never knew there were so many birds in the park. We need to make more time for this."

She appreciated the birds, but there was more to her statement than avian appreciation. Since our introduction, all of our interactions had involved alcohol. Hanging out at bars was the default for our twenty-something cohort, and every event worth attending—art show, sporting event, fundraiser, concert, or party—featured drinking to some degree.

Sonia always prioritized people over partying. Drunkenness was never her goal, and she was happy to nurse two drinks through a three-hour conversation. Momentum occasionally overtook her, but she remained gracious throughout those instances. She never compromised her behavior or demanded attention while intoxicated, and Drunk Sonia was as attentive, caring, and lovable as the sober original.

I was a different story because alcohol augmented my outgoing personality and made me an overbearing presence at every function I attended. My desire for the next drink marginalized meaningful interaction, and I became boisterous and distracting as my intoxication steepened. I was never belligerent or abrasive, but my self-absorption and desire for audience rendered me oblivious to Sonia's needs, feelings, and frustrations.

Our Central Park outing represented a departure from that pattern and offered Sonia an extended view of what lay behind my alcoholic cloak. Absent Olde English and Jägermeister, my attention was focused on her, our interaction meandering between revealing conversation and birding interludes as we wound through Sheep Meadow and toward the park's southern edge.

Exiting the park, we ducked into the FAO Schwarz toy store, where our subsequent silliness peaked when I climbed into a bin of stuffed hippopotamuses.

Smiling, Sonia threw me a stuffed snake as a rescue rope and pulled me to safety.

I didn't think much of our tomfoolery, but Sonia returned to the store, purchased one of the hippos, and gifted it to me for my birthday a few days later. I clutched the plush animal and declared, "I'm gonna call her Yvette" (Sonia's middle name).

She gave me a death stare. "Great," she said. "Just what every woman wants—a hippo named in her honor."

I countered, "We could exchange it for that big crab if you think it suits you better."

She rolled her eyes and kissed me; she knew it was the only way to shut me up.

In late January, to return the hippo gesture, I bought Sonia a stuffed seal for her twenty-ninth birthday. Lending the animal my voice, it barked out an off-key rendition of "Happy Birthday."

"This is the best gift ever!" she said with a beaming smile.

We enjoyed a quiet birthday dinner before meeting a bunch of Sonia's friends at a bar in our neighborhood. The early part of that celebration unfolded without incident, but I became rowdy as my drinking accelerated. My focus and behavior devolved, and, as midnight approached,

I removed my shirt to demonstrate how much fun I was having. Sonia put me in a cab shortly thereafter, and I had no recollection of my exit when we spoke the following day.

She lamented, "I wish you'd let my friends see the wonderful, caring man you showed me in the park and at FAO Schwarz."

I suddenly regretted opening the window to myself. I replied, "But that's just for you. I'm not going to bring stuffed animals to a bar."

Sonia's arms fell flat at her sides. "What am I supposed to do with that, Dorian?" she asked. "My friends are warning me to rethink this, and I'm wondering if I should."

I was also confused about our trajectory. Though gaga over Sonia, I never envisioned myself in a committed, long-term relationship; the bachelor life promised a continued string of sexual partners, and I harbored zero desire to be part of a nuclear family, children the stuff of nightmares as far as I was concerned. I wasn't keen on tempering my actions to fit her expectations, and I often resented Sonia as a brake on my freewheeling, late-night lifestyle. She detested the thumping clubs and seedy parties I frequented, and I was tired of compartmentalizing the dichotomous facets of my life. I was confident I could juggle science and substance abuse, but I wasn't sure I could manage Sonia as well. Something would need to give. Four months into our relationship, I didn't know what it would be.

———————

Pedaling along the shoulder of Interstate 84, I marveled that the man from my memory and the man on the bicycle were the same person. I claimed four-and-a-half years of sobriety at that moment, but I was still working to understand my alcoholism. For more than a decade, my affliction had been a ruthless governor; it occluded opportunities, prevented introspection, and isolated me from the people who would have supported me if I'd chosen to confront my condition. Those costs were obvious in sober hindsight but ignored in the moment; like the python on which Sonia's rescue rope was modeled, my alcoholism had constricted my perspective. At that four-month point in our relationship, neither of us knew if she'd be able to pull me toward sobriety.

Lost in my thoughts, I cranked out the remaining sixty miles to Le Grande and powered into southern Washington via Tollgate the following day. Overnights in Walla Walla, Pasco, and Ephrata marked my course north, and the undeveloped stretches between those points gave me a wonderful sense of Central Washington, sagebrush mesas and rocky ridges prevailing as I continued north. By the time I reached the upper throws of the Columbia River on August 15th, I'd cycled 475 miles in the six days since I departed Boise.

I met biologist Michael Schroeder at his Bridgeport home that afternoon. Working for the Washington Department of Fish and Wildlife, Michael has dedicated his professional life to advancing our understanding of grouse biology and distribution. To say Michael had a fondness for grouse would be the grossest of understatements; he might like grouse more than I once liked cocaine. He was wearing a prairie-chicken T-shirt when I arrived. The walls of his modest abode were plastered with images of various game birds. His internet network was named "Grouse." His dog was called "Lek." If grouse was a drug, then he'd use it by the gram. It was inspiring to see obsession deployed so productively.

Strategizing over a table-sized map of northern Washington after dinner, Michael pointed me toward Conconully, a 200-person village on the eastern slope of the North Cascades. High above town and only twenty-five miles from the Canadian border, I'd find a swath of spruce forest known to harbor the tender Boreal Chickadee and the shifty Spruce Grouse. Those species were the centerpieces of my Washington extension, so I was prepared to invest several days into the search for them.

I thanked Michael for his hospitality and guidance and set out on August 16th. The first fifty miles of that ride were straightforward; accompanied by a modest 2,000 feet of climbing, the distance was behind me before lunch. The final 20 miles, however, required 4,200 feet of uninterrupted gain. Relentless cranking earned me glimpses of a brownish Boreal Chickadee toward the top of the ascent.

The following morning, I began a more focused search for the Spruce Grouse, a species that survives the boreal winter almost entirely on spruce needles. It took several hours of hiking and bushwhacking, but I eventually intersected a male lurking in the underbrush. The pigeon-sized bird permitted close approach, and I had mesmerizing views of his blue-gray plumage as he picked his way through the low-slung, coniferous labyrinth. He eventually tired of my bumbling pursuit—I was on hands and knees for much of it—and flew up into a tree. I left him to roost, extracted myself from the forest, and descended toward Okanogan.

I was ecstatic that I'd found the chickadee and the grouse, those intersections validating my Washington wanderings, but I wondered how I would have felt if I'd missed either or both birds after investing what could have been several additional days. Disappointment and frustration were my reflexive answers, but I realized I had limited control over any birding outcome; no amount of research or riding could guarantee I'd observe any species beyond the most common. Chance permeates birding, and the game of hide-and-seek would be boring without it; like gambling or sports, winning is only valuable against the threat of losing. It's that uncertainty that makes the Big Year game worth playing.

Speeding downhill, I realized I had more control over decisions than outcomes because the latter is subject to the confounding influence of randomness over time. I was fortunate to find the chickadee and the grouse, but there were countless hypotheticals that would have caused me to miss one or both birds. Maybe I tore a calf muscle on the climb to Conconully; perhaps a hospitalized parent necessitated an emergency trip to Philadelphia. More likely, I simply looked in the wrong direction when either bird emerged from the forest.

I suddenly recalled the serenity prayer. Recited while holding hands at the end of every Alcoholics Anonymous meeting, the collective appeal acknowledges the immense gap between what a person can control and what they cannot. Alcoholism and addiction thrive when afflicted parties feel powerless, so the closing ritual empowers attendees to claim agency in the simplest terms: stay sober today and—hopefully—reaffirm that commitment tomorrow. Absent the ability to control events beyond our present decisions, alcoholics, transcontinental bike-birders, and everyone in between mustn't let uncertainty or fear of failure dissuade them from changing course or undertaking difficult journeys. Agency—like a leg muscle—is useless unless it's flexed. Once I decided which direction to point the bicycle, all I could do was pedal and hope subsequent events broke in my favor. As long as I reaffirmed my coincident daily commitments to sobriety and cycling, I trusted I'd find something valuable along those intertwined arcs, that affirmation much more important than the chickadee or the grouse.

TWENTY-TWO

Burned

Smoke suggested fire as I crested Loup Loup Pass, and odors of charcoal invaded my nostrils as I descended into Washington's Methow Valley. Wheeling through a ghostly grove of pines, their needles bleached gray by the inferno, I entered a smoldering graveyard. The understory was obliterated, and hundreds of thousands of blackened trunks occupied the surrounding mountainsides. My spirits sank as I rolled deeper into the devastation. No animals scampered. No birds sang. No insects buzzed. If there was a highway to hell, then I was biking it.

I'd cycled through burned swaths during my Rocky Mountain summer but hadn't confronted active combustion until I steered into what remained of the Carlton Complex wildfire. That historic blaze began on July 14th as four distinct fires, each ignited by lightning. Extended drought, elevated temperatures, and ferocious winds merged those quaternary events into a single conflagration over four days, and the resulting behemoth forced evacuations and incinerated several hundred homes as it raced across the Methow Valley. Encompassing 240,000 acres at 2 percent containment on July 20th, the wildfire had grown into the largest in Washington's history in just six days.

Diminished winds, soaking rains, and heroic firefighting rendered the blaze 90 percent contained by August 10th, the day I initiated my Washington extension, and I monitored the situation as I approached Conconully through the next week. By the time I secured Boreal

Chickadee and Spruce Grouse, the fire was all but extinguished and Highway 20, my envisioned route west, was reopened. I spent two nights in Okanogan post-grouse sighting, the intervening day used to add White-headed Woodpecker as species #520, and overcame Loup Loup before descending into the carbonized Methow on August 19th.

I intersected a meandering river at the bottom of the valley and tacked northwest toward Twisp, that course setting me straight into the Methow's prevailing gales. Those billowing breezes breathed life into the Carlton Complex fire a month earlier but had the opposite effect on me, the airflow sapping my strength and fanning my frustration as I forced my bicycle forward.

What did I do to deserve this? Can't anything be easy?

The wind hissed in my ears, smacked my face, and pushed me about the road; a forceful gust sent me into the traffic lane. I narrowly avoided colliding with an overtaking vehicle, and I nearly lost my balance trying to regain the shoulder. The blustery beatdown was wringing me of motivation and patience. Another rush of air sent me into the guardrail. The impact dissipated my momentum, and I unclipped my feet before the bicycle collapsed. I stumbled forward and reclaimed balance by scraping a palm along the asphalt.

Fuming, I grabbed my bicycle by the frame, hoisted it to my chest, and heaved it into the roadside ditch before storming away and collapsing on the grass. Self-doubt overtook me, and I wondered if my Washington plans were too ambitious, the remaining twenty miles to Mazama an impossibility from that wind-whipped vantage. Legs listless and spirit suffocated, I stared at the charred hillsides, looking for answers.

———————

As I was saying goodbye, I blurted, "I love you."

I hadn't planned the declaration; it just happened, the way an unattended pot of pasta boils over. With Sonia in Toronto and me in New York, a phone call was the least romantic way to express that sentiment for the first time.

"Thank you," she said after an awkward pause. "I'll talk to you tomorrow."

Though she didn't echo my sentiment, I wasn't taken aback. We'd been together for only five months, and I didn't know how to profess my feelings because I had zero relevant dating experience. Then thirty years old, my longest tenure with one woman at that point was a six-month, cocaine-fueled hook-up with a twenty-one-year-old when I was twenty-seven. I'd long confused intimacy and intoxication, and I hadn't so much as kissed a woman without the aid of alcohol since high school, Sonia recently excepted. While she wasn't privy to those particulars, she guarded against my obvious emotional immaturity and expressed a mix of frustration, disappointment, and curiosity whenever I had ten drinks too many. Try as she did, she couldn't fathom the power alcohol held over me.

"Why do you do this to yourself? Why am I not enough for you?" she asked a week after her return from Toronto, the morning after another rowdy night and accompanying blackout.

I did my best to answer her, but explaining my unquenchable thirst for alcohol to someone who wasn't similarly afflicted felt like describing Queen's "Bohemian Rhapsody" to someone born deaf.

Despite that disconnect, Sonia didn't place restrictions on me. It wasn't her responsibility to force behavioral change, and we both knew I'd resist governance however kindly it was couched. She'd watched friends and coworkers move beyond drinking as they'd found life partners and made professional advances, and she expressed hope I'd experience a similar slowdown. A timetable for that maturation never discussed, Sonia waited for a taper that my condition would never allow.

Against alcohol's ambiguities, drugs presented clarity. Given my DJ hustle and affinity for all-night clubs and parties, Sonia surmised I had a narcotic history. She wasn't concerned with what I falsely led her to believe were punctuated and past indiscretions, the ongoing reality worse than she could have imagined, but she made it clear that continued drug use was a deal-breaker. She didn't state her position in an accusatory or pejorative tone; rather, she revealed it in casual conversation, the way I might remark I could never date a woman who hunted ducks, denied evolution, thought climate change a hoax, or rooted for the New York Yankees.

Sonia was enchanting, so I never thought about drugging in her presence. She did, however, frequently travel for work, and those absentee stretches allowed me the binges I craved. I needed a break by the time she returned from a weeklong trip, and I believed I could extend my deception so long as I could restrict drug use to periods when she was out of town.

I couldn't, and we experienced our first serious conflict a few weeks after I first professed my love. On that Friday night, Sonia and I joined my NYU friends at a local dive. Initially engaging and enthusiastic, Sonia faded as midnight approached.

"I'm going to head home," she said. "But you should stay and hang out. I'm having brunch with Fernanda tomorrow, so we can meet up in the afternoon."

I walked her to the door and gave her a hug and kiss. "Get home safe," I said.

"Will do," she said. "I'm going to walk since I need the air. See you tomorrow."

Sonia strode away, and I ducked back into the bar. Three more beers and the itch for cocaine overtook me. I hadn't scratched since Sonia returned from Toronto, so I called my dealer and asked him to be at my apartment in thirty minutes. I pounded another beer, called a few cronies, and grabbed a cab with my friend Melinda, a redhead who looked more supermodel than scientist in her skin-tight dress. Melinda and I took delivery of the drugs at my place, six others joined us an hour later, and we partied until the cocaine gave out. I slugged two final beers as a sleep aid and climbed into bed at six a.m. With a few hours of sleep, I'd be fine by the time I met Sonia.

I awoke around noon. Sonia called shortly thereafter.

"How'd your night end?" she asked.

I replied, "Great. Stayed at the bar until three. Walked home, grabbed a slice from Artichoke [the popular pizza place directly across the street from my apartment], and crashed. You?"

"The walk home was nice," she said. "Weird thing, though. I saw you and that redhead going into your apartment while I was waiting in line at Artichoke. What was that about?"

My diaphragm stalled, and my head spun as I tried to concoct an explanation to clear myself on the infidelity insinuation while concealing the cocaine. Unable to do that against sudden hypoxia, I disclosed the drugs and the associated gathering.

"I thought you were done with that crap!" Sonia said forcefully. "You know how I feel about drugs. You snuck around and then you lied about it. I don't want to be involved with someone I can't trust."

She hung up. I spent the weekend apologizing through voicemails and text messages. My pleas for forgiveness went unanswered through that night and the next, but her position softened on Monday, when she agreed to meet me at my apartment.

Seated on my sofa, Sonia explained. "You really hurt me, Dorian. I don't want to trash our relationship over it, but I need to know you're done with drugs," she said.

"I'm done. I promise," I replied. "It was an isolated meltdown. I'm sorry. And I know lying was just as bad. It won't happen again."

"If it does, I'm out. Understand?" she reiterated.

I had no idea if I would honor my pledge; I only knew Sonia needed to hear it.

We moved beyond my transgression in subsequent weeks. I abstained from drugs through the remainder of the spring but backslid in June while Sonia was traveling for work. When Sonia went to Rochester to visit a friend in August, I continued my behavior, a drug-fueled blitz delivering me to an all-night club in the West Village and an all-morning loft party in Brooklyn. That after-hours event winding down at noon, I took the subway back into Manhattan.

Sonia called shortly after I stumbled into my apartment. Still under the combined influence of alcohol, ecstasy, cocaine, and marijuana, I made the overconfident mistake of answering the phone. She knew something was wrong and pressed me on inconsistencies I couldn't square. My series of lies was crumbling, so I confessed the truth with the hope that my admission would garner leniency.

It didn't.

"I can't do this anymore. I'm done waiting for you to get your shit together. We are done. Don't bother calling."

I was furious with myself for answering the phone, but the lingering effects of narcotics prevented deeper reflection. Defaulting to denial, I decided a breakup was best because then I could do what I wanted without hurting Sonia.

My perspective evolved in subsequent days, time and sobriety suggesting deeper remorse for the disrespect I'd shown Sonia. I busied myself in the lab to distract myself from growing regret but left her a short voice message the following week. I didn't suggest plans to address my drinking or drugging or hint at forgiveness; I only expressed the wish that she take care of herself and find someone deserving of her boundless affection. She deserved that much after enduring me.

I was shellshocked when Sonia called a week later. Though I'd assumed we were over, she said she needed more time to think. She didn't articulate specifics, but I listened and made myself available for future communication if she wanted it. Exercising that option, she called and suggested a meeting in Tompkins Square Park.

I walked to the park and claimed a shaded bench. Sonia arrived a few minutes later and, after awkward pleasantries, described the emotional terror I wrought on her. "I trusted you. I had no idea my soul could ache the way it has for the last week," she said.

I could hardly look at her, the ground a more comfortable view because I knew I was too selfish to feel anything equivalent. I replied tenderly, "I'm sorry, Sonia. I wish there was something I could do to make it better. I never meant to hurt you."

"But you did," she said. "Your decisions have consequences beyond you. Don't you get that?"

"I know. And I'm out of excuses. I can't stop fucking up." All I wanted to do at that moment was crawl into the nearest sewer. I couldn't find better words, so I asked, "You know I love you, right?"

"Goddamn it, D! Yes. And I love you!" she said with exasperation. "That's why I'm here. If you were some big asshole, this would be easy. But you're not. You're an intelligent, funny, and caring man who gets drunk and makes terrible decisions. I don't know what to do with you."

"So where does that leave us?" I asked.

She sighed. "I don't know."

I understood but didn't know how to respond. Trying to combat swelling silence, I wondered aloud, "Maybe I'll look into alcohol counseling or something."

That suggestion struck Sonia, and she turned toward me with wide eyes, as though never expecting that concession. She replied from behind building tears, "Would you really do that?"

"Yes," I said softly. "I'm not sure where to start, but I'll see what I can find."

That was next-level bullshit. I knew exactly where to start: Alcoholics Anonymous, my first stint in the program concealed across our ten-month association. I'd also had two formal counseling sessions at the Center for Motivation and Change in Manhattan—when and what circumstances drove me there I cannot recall—but I knew exactly where to find support if I wanted it.

Sonia relaxed at the thought of me getting help. She avoided references to reunification in the wake of my concession, and we departed Tompkins Square Park with only an agreement to speak in a week. Neither of us knew where renewed communication would lead, but I was encouraged she was willing to engage me.

Given harmonious personalities, shared interests, comparable professional aspirations, aligned politics, and no procreative desires on either side, my drinking—and derivative drugging and dishonesty—was the only obstacle to a sustainable and loving relationship with Sonia. If I'd decided to get sober on that park bench, then I could have all but solidified that vision. Afraid of a future without alcohol, I offered the hypothetical counseling concession instead, even with Sonia's teary eyes staring into mine. But that's the nature of alcoholism; it muddies decisions, erodes relationships, and expands into the remaining void. It didn't matter how I felt about Sonia; drinking was still the love of my life.

Sonia and I spoke the week after meeting in Tompkins Square Park and slowly rebuilt our relationship through September. I controlled my drinking through those weeks, but I never pursued the counseling as I had promised. By the time the Phillies faced the Yankees in the 2009 World Series at the end of October, the Fall Classic marking one year since we met, I'd reestablished my drinking patterns, renewed blackouts rendering me a ticking bomb again. No timeline for explosion, we plodded naively forward.

There was, however, a definite schedule for my academic advance. Beginning my sixth and final year of graduate school, I was committed to becoming a biomedical researcher and university professor. A postdoctoral fellowship was the next step, and I'd whittled my decision down to two labs: one at Stanford and one at Massachusetts General Hospital. I knew what to expect from each institution and setting, given my experience in Palo Alto and Cambridge, so my decision was based on Sonia as much as scientific considerations. She'd attended Gordon College north of Boston and always spoke longingly of New England, so I committed to Mass General on the hope that she would accompany me to Boston in November of 2010, my fellowship scheduled to start in January of 2011. In the intervening year, I could finish an important panel of experiments, write and defend my thesis, and work on our relationship.

We traveled to visit our respective families over Christmas and reunited in New York on December 31st. New Year's Eve is the most overhyped and overpriced holiday on the calendar, so I was happy to spend it on the sofa with Sonia. A week later, she spent the night of Friday, January 8th, at my apartment ahead of a morning outing to Central Park. Unable to pry herself from a warm bed when the alarm sounded, Sonia encouraged me to go without her. Birding didn't beckon often, and we both knew I needed to seize the opportunity.

As I was leaving, Sonia asked, "Can I borrow your phone to call my sister? Mine's dead and my charger's at home."

I handed her the device. "Sure." I gave her a kiss and hustled toward the subway.

The park was quiet at eight a.m., and I had nice views of wintering hawks, finches, and ducks as I explored the Reservoir, the Ramble, and the Lake. Absent my usual Saturday morning hangover, it was a rare moment of birding bliss. The crowds increased into mid-morning, so I retreated toward my apartment around eleven a.m.

Opening the door, I found Sonia waiting on the sofa. She sat up, extended her arm, and showed me a text I'd apparently sent. It read, "Sonia is out of town. Snow [cocaine] forecasted. Swing by to party."

Frantically accessing the recesses of my brain, I remembered I'd returned to New York on December 27th, four days ahead of her. I got shit-faced at Professor Thom's that night, and my heart sank when I realized I must have sent the damning text in a blacked-out stupor. I rarely texted in those early texting days, so the message was at the top of my history; all Sonia needed to do was mistakenly enter that menu on my unfamiliar device.

I didn't bother explaining that I never took delivery of the referenced drug. My intent was clear. I didn't say a word. Neither did Sonia. She dropped my phone on the floor, put on her shoes and coat, and walked out.

———

The incinerated Methow Valley finally spoke to me; though annihilated, the ecosystem would recover. Wildfire is a vital part of the forest life cycle; it purges accumulated matter, enriches the underlying soil, and facilitates growth. Grasses and wildflowers would emerge from charred mountainsides, and rejuvenated meadows would cradle upstart saplings. Insect pollinators would aid the rehabilitation, birds and mammals would repopulate the landscape, and trees would again reach for the Washington sky. From my windswept vantage, I realized that building a forest is no different from riding a bicycle into the wind: they're both slow but surmountable tasks.

Rallying to the challenge before me, I relinquished my roadside perch, dragged my bicycle out of the brambles, and resumed my upwind course. The riding was agonizing, every crank contentious as the gales shoved my bicycle about the blacktop, but I kept my cool. I'd overcome worse in my life, and the universe would need weapons beyond wind to defeat me.

Relying on my lowest gears, I slogged through Winthrop and fought toward Mazama, a destination I reached in the early evening. Climbing into bed after eating dinner and blogging, I had little time to savor the day's victory. Tomorrow I'd return to the mountains, the North Cascades looming as the next in a year-long string of obstacles.

TWENTY-THREE

Come as You Are

The Cascades extend from southern British Columbia through Washington and Oregon before terminating at Mount Lassen in Northern California. The range was formed by the slow shifting of Earth's tectonic plates, the mountains rising when the Juan de Fuca Plate slid east and under the North American Plate forty million years ago. Half as old as the Rockies and less than a tenth the age of the Appalachians, the Cascades are a geologic infant. Erosion has yet to blunt craggy peaks, and recent glaciation has deepened intervening valleys. That topographical juxtaposition is most pronounced in northern Washington, and those reaches of the North Cascades are considered the most rugged mountains in the lower 48 states. It's for that reason that the range has generally resisted development.

I'd sparred with the eastern reaches of the North Cascades at Conconully and Loup Loup Pass, but I needed to cross their north–south spine to reach the Puget Sound and Pacific Ocean to their west. With no lodging en route, I'd need to complete the eighty-mile traverse in a single day. That uphill task began as I limped out of Mazama on the morning of August 20th, the day after my Methow meltdown.

Unlike my previous climbs, which featured convoluted courses and steep switchbacks, Highway 20 ascended along the lateral slope of an immense glacial valley. A rushing river was audible below my course, and beautiful boreal forest blanketed the adjacent mountainsides as I gained elevation across the next two and a half hours. Minus the two-lane strip of tarmac I occupied, no development was evident, the North Cascades an enduring bastion of wilderness.

After 20 miles and 3,400 feet of vertical gain, I turned into the rocky confines of Washington Pass. I knew the overhead pinnacles had held for hundreds of thousands of years, but it seemed like the granite batholiths could topple onto me at any moment. As I crested at 5,476 feet, I contemplated my next move.

With Boreal Chickadee and Spruce Grouse handled and Mew Gull expected when I reached the Puget Sound, Gray-crowned Rosy-Finch, a relative of the Brown-capped variety I observed on Mount Bierstadt in Colorado, dominated my thinking. My plan was to seek this alpine specialist on Mount Rainier's glaciated slopes in a week's time, after I crossed the North Cascades and commuted through Seattle, and I'd blogged about that intention while approaching Mazama. Readying for bed that night, I received a call from birding buddy Barry Lyon, whom I met when he served as my counselor at Camp Chiricahua.

"I just caught up on your recent entries, and I have an idea," he said. "You should look for Gray-crowned Rosy-Finch at Maple Pass tomorrow. You're going to go right by the trailhead after you go over Washington Pass."

I understood his insinuation; if I could find the bird in the North Cascades, amidst elevation I'd already gained, then I could skip Rainier and go straight to the coast from Seattle.

I replied, "Gotcha. But I haven't heard of recent rosy-finch reports from Maple."

"That's cuz it's under-birded," Barry explained. "It's a tough hike, so everyone drives up Rainier instead. But the habitat is perfect, and you're in better shape than any birder on Earth."

"Interesting idea," I said. "It's prolly gonna be a game-time decision based on how I feel tomorrow, but I'll let you know what happens in the next blog entry. Cheers!"

A bit of follow-up research brought the decision into quantitative focus; the hike to Maple Pass would be a 7-mile loop and include 2,100 feet of vertical gain on the outgoing leg. It would be a chore if undertaken in the middle of my Cascades traverse—that task requiring 6,000 feet of cumulative climbing spread across 80 miles—but I couldn't ignore the possibility because intersecting Gray-crowned Rosy-Finch at Maple would save me the three days, 120 miles of riding, and 8,000 feet of climbing to reach Rainier.

I'm already beaten down. Am I willing to absorb additional and immediate pain on what feels like a long-shot chance of granting myself Rainier relief?

I was unsure of the answer by the time I fell asleep in Mazama, but I was ready to gamble by the time I reached Washington Pass. My legs survived the morning ascent better than I expected, and clear skies would facilitate afternoon time above tree line. I'd reaped rewards every time I'd extended myself, and I hoped the trend would continue. Pulling off Highway 20 at the Maple Pass trailhead, I ditched my bicycle in the woods, laced up my hiking boots, and initiated my ascent.

The lower portion of the climb wound through lush forest, the western slope of the North Cascades wetter than the rain-shadowed eastern, and I emerged from old-growth conifers on a rocky hillside overlooking an emerald lake. I was clear of trees by 6,000 feet and absorbed gorgeous alpine views as I achieved an overhead ridge and followed a precarious knife's edge toward the pass at 7,000 feet.*

Achieving that vantage on wobbly legs, I scanned the surrounding slopes and adorning snowfields for rosy-finches. My efforts felt token, so I was surprised when I heard buzzy calls an hour into my vigil. Peering downslope, I spotted five small birds sweeping up the mountainside. I didn't have time to reach for my binoculars, but the characteristic calls had already betrayed their identity. The quintet of plump brown birds flew over my head before they disappeared down the backside of the pass.

I clenched my fists and emitted a triumphant roar, the echo rejuvenating as it bounced between mountaintops. Against what felt like impossible odds, I'd bet on Maple Pass and won, my victory a potent

---

* Tree line varies as a function of latitude. Whereas 11,000 feet were required to reach the alpine tundra in Colorado, I needed to achieve only 6,000 to reach equivalent habitat in the more northern Washington.

reminder that extending myself in the face of challenge was the correct decision. Absolved of Rainier responsibility, I lolled at the summit, my hope that the birds would return for another view and, ideally, a photograph. When their absence extended to an hour, I departed Maple and descended to my bicycle. With fifty-two downhill miles to reach my Marblemount destination, I felt invincible as I rejoined Highway 20 west.

A heavy push of eastbound air extinguished my optimism in the next twenty minutes. As the Methow Valley warms through the morning, rising hot air creates a barometric depression; the resulting vacuum subsequently sucks heavier marine air over the mountains through the afternoon. That eastern push, the same that precipitated yesterday's Methow meltdown, hit me like a freight train and negated gravity's assistance on all but the steepest pitches. The unanticipated impediment was an awful insult, my legs aching as I forced the bicycle downhill.

A segment I thought would take two hours instead consumed four, and I didn't reach Marblemount until seven thirty p.m., thirteen hours after I departed Mazama. Minus the relaxed hour at Maple Pass post-rosy-finch, I'd been biking or hiking for twelve of those. I could barely stand to shower or stay awake to eat, and I conceded the entire next day as rest, that recovery dominated by crime dramas and reality television. With SportsCenter and Jeopardy! in the mix, I was master of my couch domain.

Reflecting on my high-elevation summer during commercial breaks, I couldn't believe I'd survived the mountains. Hard riding and associated exhaustion necessitated periodic rest, Marblemount being the latest example, but my legs delivered every time I needed them. I'd averaged fifty-five miles a day since departing Phoenix, Arizona, on June 1st, and I'd overcome some of the continent's most notorious ascents at Molas Pass, Red Mountain Pass, Monarch Pass, Loveland Pass, Guanella Pass, Teton Pass, and Washington Pass. Physical breakdown felt inevitable at many points, but my body—exactly as it did during my drinking and drugging days—absorbed whatever abuse I heaped on it. With 11,000 miles behind me and another 6,000 ahead, I hoped my health would hold as I moved south along the Pacific Coast through September and October.

I wheeled through Everett, Seattle, Puyallup, and Olympia in subsequent days. Mew Gull presented on Puget Sound as expected, and I added a surprise Slaty-backed Gull, a vagrant from the Asian Palearctic, in Tacoma. That bonus secured as species #525, I steered toward Aberdeen, the blue-collar birthplace of Nirvana front man Kurt Cobain.

I'd reached out to a local cyclist about lodging there but had not heard back by the time I rolled into town on August 28th. I pulled into a gas station, bought a hot dog, sent my contact a text message, and doused my elongated snack in ketchup and mustard while awaiting his reply.

He texted a moment later: "Sry for slow reply."

I replied, "No prob. U cool to host tonight? Chillin n town now."

"Working til 9 but u can swing by then."

The late arrival was fine. I could enjoy some afternoon birding before hunkering down at a fast-food place near his house. Between eating dinner, researching routes, and blogging, the hours would evaporate.

"Ok thx c u then," I texted back.

Lodging secured, I addressed my hot dog. Another text came through as I bit into it.

"Forgot to say I'm nudist."

I nearly choked, my foot-long losing all appeal in that instant. Though shocked, I couldn't control my curiosity. Getting out of my comfort zone had lent education and inspiration at many points in my journey, and the latest invitation would provide additional opportunity to explore beyond my usual boundaries. As I weighed the pros and cons of accepting his qualified housing offer, I received another message.

"And I hope ur ok being nude 2."

I had no idea if my potential host was serious or not, but the $50 I dropped elsewhere on a rickety bed, dripping faucet, and broken television was the deal of the century. Channeling "Come as You Are," Nirvana's 1992 anthem and Aberdeen's appropriately adopted motto, I departed town exactly as I arrived: fully clothed.

I reached the Pacific at Ocean Shores the following day, August 29th. Absorbing that shimmering abyss from the jetty at the south end of town, I experienced a deep sense of accomplishment, my transcontinental journey initiated from a similar structure at the mouth of the Merrimack River in Massachusetts 241 days earlier. My Atlantic view on that frigid January morning was one of the future, my outlook steeped in inexperience, apprehension, and personal confusion. I had no idea how my adventure would unfold nor any clue how I would survive the journey; I felt only an indescribable motivation to board the bicycle and search for birds.

In the intervening eight months, I'd mastered long-distance cycling, experienced an incredible array of American landscapes, overcome incalculable challenges, and found 530 bird species toward my goal of 600. My Pacific view was thus one of the past, a personal recollection of the physical and emotional distance I'd traveled since my Atlantic departure.

My retrospection also recalled my life prior to my bicycle Big Year. I was proud of most, ashamed of some, and regretted none; if I'd done anything differently, then I would not have experienced the perspective my jetty view permitted. Each decision or event influences the next, sometimes to unexpected ends, and randomness guarantees it's impossible to know where an alternative path would have led were any option exercised differently. Unlike the rise and fall of tides, the human experience isn't calculable; it's unapologetically unpredictable and proceeds without pause. Standing next to my bicycle, staring at the Pacific, the alternatives and hypotheticals didn't matter; my past was penned, the present was perfect, and the future was wide open.

Meditation aside, new birds abounded as I scanned the ocean. Acrobatic Elegant Terns wheeled like overhead angels as they pursued baitfish in the inlet, greedy Western Gulls approached with hopes of handouts, and a Wandering Tattler, a sandpiper that prefers rocks to beaches, scoured the jetty's waterline for invertebrate morsels. I'd observed only forty-two new species across the previous two months, so adding ten in two hours on the jetty was a fitting reward for reaching the Pacific, albeit along the most convoluted route imaginable.

That injection of birds was welcome, but all ten species were expected. More critical was Pacific Golden-Plover, a Siberian-nesting shorebird that migrates to Southeast Asia, Australia, and Oceania for the nonbreeding months. A few representatives cross the Bering Sea and sojourn on the American side of the Pacific each winter, and the bird is a rare-but-regular visitor along the West Coast between August and April. There were several recent sightings from Ocean Shores and

Westport, but I was unable to locate a representative despite investing two and a half days in and around those municipalities.

Confident I could redeem my plover miss as I moved south, I rolled out of Westport and followed Highway 105 onto the northern edge of Willapa Bay. Evergreens overhung the left side of the road while waves lapped at the right, and I savored complementary odors of sweet pine and salt water as I rolled through Raymond, joined Highway 101, and skirted the bay's southern section.

Powering beyond Willapa, I intersected the mighty Columbia River and followed the wide flow west to the Astoria–Megler Bridge, the longest truss bridge in North America. Bicycles are permitted on the four-mile span, but a skinny shoulder and oversized vehicles created the tightest riding I'd experienced. Massive RVs were particularly hair-raising; with six inches between them and me, I thought I was going to be crushed against the bridge railing. I survived the elevated cantilever section at the span's southern end, touched down in Oregon, and struggled south to Cannon Beach, 108 miles from my Westport starting point.

A charming coastal retreat, Cannon Beach represented the end of my Washington extension. Besides adding 700 miles to my original route and netting five species I would not have found otherwise, the undertaking delayed my accelerated course so it better aligned with migration; had I reached Cannon Beach on August 20th instead of September 1st, I would have moved south ahead of some of the birds I needed to find on my autumn leg. Assuming I didn't burn out before I completed the remainder of my route, the Washington extension looked like a wise stroke. As with most decisions, only time would tell.

I hobbled to the town beach. The sandy expanse stretched a mile in each direction, and masses of beachgoers absorbed views of Haystack Rock, a 235-foot pyramidal chunk of volcanic basalt rising straight out of the surf. Though the town's geological signature can be approached at low tide, Haystack's steep faces prevent ascent, its upper throws offering cordoned refuge for gulls, cormorants, and alcids.

Among the birds that nest on Haystack, Tufted Puffins are the most recognizable. Black-and-white with huge orange bills, the portly birds are named for their bushy golden eyebrows. Their faces are tender and expressive, and the town has adopted the species as its unofficial mascot; photos and caricatures of the adorable alcids grace the doors, walls, and tables of many local businesses.

While observable in the summer as they commute between Haystack and the surrounding fishing grounds, the puffins migrate far offshore after breeding. The vast majority vacate Haystack by late August, though a few linger into early September. With nowhere to find the puffin farther south, I was stoked when a straggler appeared over the ocean, circled the rock twice, and crash-landed at the mouth of his earthen burrow. No worse for wear, he waddled into his hole as the sun dipped beneath the horizon. Species #543 checked, I limped back to my lodging and promptly passed out.

My Cannon Beach exit was spectacular. Following Highway 101 south, I alternated between rocky shorelines and driftwood-strewn beaches. Impassable cliffs forced the road into the forest near Oswald West State Park, and I gained several hundred feet of elevation before emerging from the trees. The Pacific seemed an endless blue expanse from that promontory, and the Nehalem River wound through pastures to the east of my vantage. With clear skies, warm temperatures, and minimal traffic, the riding was idyllic as I cranked toward Tillamook.

I continued to Newport, cut inland to Corvallis to add Mountain Quail, rolled through Eugene, and regained the coast at Florence before reaching Bandon on September 10th. Pacific Golden-Plover was reported on the town mudflats the two days before I arrived, but I was unable to connect with my newest nemesis at that annual haunt. My final Oregon overnight was at Gold Beach, where the Rogue River meets the Pacific. The ocean view from the porch of my motel was pristine, and I knew I'd miss understated Oregon as nightfall claimed my final day in the Beaver State. The following morning I'd begin a six-week romp through the nation's most populated and bird-rich state—California.

TWENTY-FOUR

Under the Microscope

Overhead branches kept the underlying air cool and damp, and what little sunlight reached the understory did so as discrete beams. They danced across my face as I wound among broad and gnarled trunks. Each time I thought I'd found the largest example, I was disproved around the next bend; a tree twenty feet in diameter was dwarfed by a thirty-foot example. I peered skyward to get a sense of its height, but thick canopy forced me to use my imagination. The ancient titans rendering me awestruck, I rolled deeper into their sanctuary.

Coast redwoods flourish between southern Oregon and central California, specifically where fog is available for hydration. While smaller species rely on capillary action to absorb groundwater and transport it up their trunks, that delivery method is negated by gravitational forces as trees increase in height. To supplement bottom-up hydration, redwoods have evolved the ability to absorb fog through their needles, a top-down method that facilitates growth to heights of 350 feet. Reaching weights of 800 tons (1.6 million pounds) in their 2,000-year lifespans, redwoods are among the largest and oldest organisms on Earth.

Sadly, 95 percent of old-growth trees have been cut since 1850 because redwood lumber is prized for its durability, insect repellence, and fire resistance. Much of the remaining 5 percent is protected, and Prairie Creek Redwoods State Park in Humboldt County, California, shelters some of the largest survivors. Exploring the reserve on September 13th,

I couldn't imagine the loss if those extant giants were felled. Theirs was a magnificent temple in which all denominations could worship. I soaked in the serenity, Mother Nature's silent sermon engendering a reverence no preacher could replicate.

From northern forests to southern deserts and from eastern mountains to western beaches, California presents incredible habitat diversity and hosts a corresponding assortment of birds; with approximately 680 recorded species, the state outpaces all others. The birding is excellent year-round, but fall migration is the best time to visit. My original plan was to spend most of September and all of October in California, so my arrival on September 12th was perfect. I'd have six prime weeks to explore the Golden State before turning east and returning to Texas, where I expected the Big Year clock would run out.

Cranking south from Prairie Creek, I scoured tidal estuaries around Arcata and Eureka for Pacific Golden-Plover but conceded the bothersome bird after two days, again with hopes of finding it farther south. Following Highway 101 after adding Golden-crowned Sparrow (#549) and Red-breasted Sapsucker (#550), I covered eighty miles to Benbow on September 16th and ninety to Ukiah on the 17th, that headwind-plagued ordeal featuring 4,000 feet of climbing. I pitied the Chinese buffet across the street from my motel: after I left, the restaurant declared General Tso's Chicken an endangered species.

I rolled out of Ukiah and into Sonoma County's vineyards. Even at the height of my alcoholism, I rarely drank wine. Malt liquor was cheaper, and bourgeois concerns of aromas and notes didn't interest me. Had I offered Ray, the working-class guy seated between Sonia and me at Professor Thom's on the night of our introduction, a glass of Cabernet Sauvignon instead of beer and whiskey, I doubt he would have moved so readily. Though alcohol facilitated my conversation with Sonia on that night, my abuse of it ultimately pushed her away.

———

Three days after finding my ruinous text message and fleeing my apartment, Sonia sent me an email. In a complimentary tone, she praised my intelligence, sense of humor, and capacity for kindness; using more assertive language, she highlighted the disrespect my drinking, drugging, and dishonesty had done to her and our relationship. She concluded, "I'd love for you to get sober, but that's a journey you'll have to undertake without my help. You've abused my trust, and I cannot endure more lies, broken promises, and heartache. Please take care of yourself. Goodbye, Dorian."

Neither judgmental nor bitter, Sonia's words were an accurate assessment of the distance between us, my drinking an unbridgeable chasm into which she'd fallen too many times. I possessed no ability or desire to regulate alcohol intake and—absent unconditional sobriety—I'd inevitably alienate her; that we lasted fourteen months was a miracle. Science and drinking had sustained me prior to her, and I leaned on those familiar crutches in the days following our split. Between experiments and blackouts, I didn't allow myself the emotional space to acknowledge her departure.

A week removed from Sonia's email, I met up with one of my Stanford roommates. Mike was in New York for business and arrived at my apartment after dining with colleagues. We hadn't spoken since he married and moved to Norway three years prior, so I described the incredible woman I'd loved and lost since our last communication.

Mike responded, "She sounds pretty amazing. And super laid back, you know with the sports and the outdoors."

I agreed. "Yeah, she was the opposite of high maintenance. Christ, she told me never to buy her diamonds since she thought the whole industry was a racket. And she didn't want kids. She was perfect."

"And you still managed to screw it up," Mike said with a laugh.

"Smooth, right? But I have to find someone who lets me be me. All the sneaking around was a pain in the ass, and I'm not ready to shut the party down. I have the rest of my life to be lame."

Mike interjected, "I remember you saying the same shit when we left Stanford. Nine years later and you're even more out of control."

I took a drag on my cigarette, pounded the remainder of my beer, exhaled the cached smoke, and replied, "I'm killing it in the lab, so who cares? It's getting stuffy in here. Let's bust out for a bit."

We relocated to Professor Thom's, where Mike opened a fascinating window to his life in Norway. An engineer-turned-designer, he extolled

Scandinavian minimalism, social democracy, and family life, his second child recently born. As eleven p.m. approached, he retired to his hotel.

"Take care of yourself, bro," he said. "Slowing down isn't all bad. You don't want to let another Sonia get away."

I replied, "I gotcha. It was great to catch up. Thanks for hanging."

Mike headed to the subway, and I ducked back into the bar—at least for a while. The receipts I found in my pockets the following morning indicated I spent the rest of the night in the West Village, but I didn't recall any of it.

I showered and walked to my laboratory. Still drunk, I initiated a panel of procedures before ducking into the microscope room. That darkened vault was a welcome refuge. I assumed the chair and flicked on the microscope. With the laser's internal fan emitting a soothing hum, I placed my first slide onto the stage.

My headache abated as I scored my samples, and swelling sober headspace motivated me to revisit my conversation with Mike. The exchange was the first time I'd verbalized the full chronology of my relationship to someone unfamiliar with it, and I recalled describing Sonia in glowing terms because doing otherwise was impossible. Alone in the darkness, her departure suddenly weighed heavier.

In a city of millions on a planet of billions, I realized I'd found the one woman who was capable of loving me more than I loved myself. She'd absorbed my worst while begging for my best, and she'd made a lasting relationship seem possible against an affliction doing its damnedest to prevent it. I gave her everything she wanted except an honest effort at sobriety; worse, I extinguished her desire to help me along that path, were I brave enough to walk it. I sobbed as I realized what my alcoholism had cost me. Turning the microscope on myself, the image of a broken man came into focus.

Reflecting on fourteen years of heavy drinking, I finally understood the cumulative toll alcohol had extracted. I'd cut myself on broken glass, fallen down stairs, and tumbled off porches, my worst accident suffered at Harvard. Departing a house party after consuming what I was later told was a dozen beers and near-fifth of vodka, I took a face-first fall onto the sidewalk. Friends escorted me home, but the orbital contusion left me with a bloody and swollen face. Headaches kept me out of work for a week, but I was back at the bottle as soon as I was able.

I'd survived academia but compromised achievement at every level, most notably at Stanford, where I spent more time drinking than studying. I was productive at Harvard but regularly missed days on account of alcohol, my sidewalk face-plant the most glaring example. My graduate interview at UCLA was a disaster; I got smashed at an evening social event, fell off a bar stool, and shattered several glasses. I behaved similarly during my visit to the University of Washington, and neither school offered me admission. Even after I landed at NYU, my drinking caused friction with my advisor.

I was arrested for drunk and disorderly conduct at a football game at Stanford during my first year and convicted of drunk driving after my third. A six-month license suspension taught me little, and I reverted to the dangerous and criminal practice at periodic intervals in the ensuing years, fortunately without incident or apprehension. I experienced additional legal trouble at NYU, when I staggered into Lexington Avenue at three a.m. I was nearly struck by a car, and I was again charged with drunk and disorderly conduct when I kicked the driver's door without noticing a nearby police car. Despite flirting with death and being cited, I was drinking the next night.

Recalling those episodes and others, I realized I was lucky to have survived without compromising my health, destroying my career, ruining my finances, killing someone, or suffering incarceration. Losing Sonia was gut-wrenching, but I realized a bruised heart would be the least of my problems if I extended my self-destructive behavior. My return to Alcoholics Anonymous that night—January 19th, 2010—wasn't a grand emotional metamorphosis; it was calculated insurance against what I finally realized was impending calamity.

I reestablished my evening rhythm at the Mustard Seed AA group and minimized social contact by staying busy in the laboratory. I hoped time would grant me the tools and confidence to approach sobriety differently, but obsessing over my experiments felt like my best recourse at the outset. I knew sobriety would be a protracted process, so I didn't want to force it. As long as my commitment to a sober future was intact, I was headed in the right direction.

In contrast with my first stint in the program, that reasoning was an important distinction. That first turn on the AA carousel was motivated by guilt, and I viewed sobriety as a jail sentence—miserable but temporary. Once I paid my penance, I'd be released with a clean slate and an imagined ability to regulate alcohol intake. As the only voting representative on the parole board, I decided thirty-eight days of sobriety constituted sufficient rehabilitation. The night of my release, I initiated a three-year run of recidivism, which left me exactly where I started: blacked-out and beaten down.

My second turn felt different because I knew I had to make my sobriety stick. My previous stint suggested I'd experience depression and disorientation, but I understood now that those were unavoidable hurdles if I was to guarantee my health, career, and freedom moving forward. While Sonia's departure forced the necessary reflection, I had zero designs on winning her back; trying to undo the past would only hamper my ability to move forward. If she proved to be the price of self-preservation, then I was ready to pay it.

Escaping the vineyards, I powered south toward Santa Rosa. My bicycle Big Year had afforded me thousands of solitary hours to think about my alcoholism, but I didn't know how my adventure would shape my understanding of myself. With three and a half months and 4,000 miles of riding ahead, I was confident in my course as I powered into Sebastopol on September 18th.

Continuing to Petaluma, I commenced my search for Black Rail, the continent's most elusive bird. Sparrow-sized and nocturnal, the skulker is a concealing blend of black, gray, and brown, with white flecks and red eyes lending definition to an otherwise shadowy form. I'd not encountered the species prior to initiating my bicycle Big Year, and I failed to score it once underway despite investing several late nights and early mornings in the Anahauc marshes while I was on the Texas coast. I knew I'd have additional opportunities to see at the bird, so I tabled it until I reached California.

My chances of seeing the bird in its San Francisco Bay stronghold were slim, but Big Year convention stipulated I could count it toward my total if I heard one vocalize. Most talkative when seeking mates during the spring breeding season, Black Rails will respond to recorded calls—perceived territorial intruders—at other times of the year, albeit with reduced vigor.

Armed with a battery of vocalizations on my cell phone, I ventured into Petaluma's marshes on three consecutive nights. I needed only one response, but I was unable to detect the shy bird despite its sure presence. Those overnight efforts upended my sleep schedule and rendered my eyelids heavy as I moved south to Santa Venetia on September 22nd. Only at daybreak on the 23rd did I finally hear the high-pitched *KEE-KEE-ker* call from one bird and a lower growl call from another. Though the second example was less than twenty feet from me, the figment stayed characteristically out of sight. The audio was a tremendous win regardless, and renewed optimism guided me through San Rafael and Sausalito and onto the Golden Gate Bridge by midday.

Blue skies prevailed as I wheeled across that span. Alcatraz Island was visible in the middle of the bay, and sailboats crisscrossed the surrounding currents on gathering breezes. Pastel homes peeked over each other like spectators at a sporting event, and throngs of people dotted the grassy waterfront. Compared to the other major cities I'd experienced, San Francisco was paradise.

I wheeled off the bridge and battled traffic through the city and down the peninsula, Sabine's Gull (#557) added en route to Palo Alto. Riding through the Stanford campus after dinner, I thought about the physical and emotional distance I'd traveled since graduation. I remembered chucking the Frisbee around Roble Field, doing keg stands at frat parties, working in my laboratory, and attending a few classes in between those more pressing activities. Though I envisioned postgraduate life as a blend of science and partying, the future held more than I could have predicted from my collegiate vantage point, my eventual course as concealed as a Black Rail under marshy cover. Uncertainty being life's greatest promise, I was happy to have found my way back to Stanford—however circuitous the path was.

TWENTY-FIVE

News of My Nemesis

Pescadero, California, is snuggled into the coast fifty miles south of San Francisco. Cutesy cafes welcome Silicon Valley tech-types on weekends, but the intervening Santa Cruz Mountains block out most of the high-paced hubbub, the agricultural hamlet a nostalgic throwback to the rural California communities that inspired John Steinbeck in the early twentieth century. I reached Pescadero and steered toward the home of Mark Kudrav, a local birder who'd offered to host me as I progressed south toward Santa Cruz and Monterey.

Mark received me at his cottage on the afternoon of September 24th. In his thirties and of medium height and build, he sported shaggy brown hair and a wooly beard. His deep-set eyes were framed by an old pair of wire spectacles, and I imagined him moonlighting as a local wise man outside of his hours as an early childhood educator. My guided tour of his wooded property strengthened that sentiment. Mark's knowledge of the natural world was only exceeded by his infectious appreciation for it, and his passion for birds, plants, and insects suggested this teacher was an inspiration to his students. Mild-mannered and authentic, he was an appropriate reflection of the tiny town that he called home.

A kindred spirit, Mark undertook his own bicycle Big Year the previous year, in 2013. His teaching responsibilities restricted his two-wheeled wanderings, but he explored his coastal surroundings on weekends and traced a six-week loop through California's interior during summer vacation. He ultimately covered 5,000-some miles and found 326 species, a total surpassing the 318 that Jim Royer amassed from his home base in San Luis Obispo, California, in 2010.

Mark and I spent the evening swapping stories about our adventures. We laughed about the ridiculous situations in which we found ourselves and commiserated over the common challenges we faced. His recount of his quest for Cactus Wren in the Mojave Desert was apropos.

"It was the middle of July and 110 degrees. I didn't know what the hell I was doing and ran out of water halfway through. I almost died of heat stroke, but I toughed it out and found the bird!" he said.

I replied, "Yeah, every bird is a battle on the bike, but it feels like you're earning it, ya know?"

Mark agreed, and we extended our exchange through dinner before he offered a revealing confession. He described how on January 3rd, just three days after the conclusion of his 2013 bicycle Big Year, a friend sent him a link to my blog. Mark explained his reaction after looking at it. "I thought, 'Son of a bitch! I busted my ass to get 326 species and this clown is going to try to get 600 the next year? I hope he quits in a month.'"

Mark had been friendly since I arrived, so I was surprised he'd harbored resentment. He clarified, "I'd just finished, and I was finally able to savor my accomplishment. Jim's record stood for four years, and I was having fun imagining how long mine might last. When I saw your plan, I had my answer: four months. I was pretty bummed."

I responded, "Dude, you held down a full-time job during your Big Year. I have zero responsibilities. Our projects aren't comparable. I'm in awe of what you did."

"Thanks, man. That means a lot. I just got tired of people asking, 'Do you think Dorian will break your record? How's it going to feel if he does?'"

"I get it. I'm over comparisons to Neil Hayward," I said.

A pharmaceutical executive suffering career confusion and depression, Hayward undertook an epic and healing petroleum-powered Big Year in 2013. The chummy Brit captivated the birding community as he jetted around the continent, and his final tally of 749 species topped the 748 that Sandy Komito posted in 1998.* To break the most coveted record in competitive birding, Hayward flew 194,000 miles (177 flights), drove another 52,000, and spent fifteen days at sea.†

I continued, "Several people have asked me, 'You don't seriously think you're going to break Neil's record on your bike, do you?' It's like they thought I was trying to one-up someone who could fly a thousand miles in the time it took me to pedal twenty!"

"I know!" Mark said. "That's why I was protective of my record. I wanted my 326 to be a reminder that process matters."

I replied, "And it will, right alongside Jim's 318, Ron Beck's 301, and however many I end up with. Anyone with enough money has a shot at winning the Big Year or breaking a record, but we're wired differently. As you said, it's more about how than how many."

Mark nodded. "I gotta admit, I was rooting against you for the first few months because I was tired of hearing about your project."

He recounted an episode from March, ten weeks after his Big Year finished and mine commenced. A driver spotted Mark bike-birding along the Pacific Coast Highway and pulled over to engage him. The driver thought Mark was me—"Are you the Biking for Birds guy?"—but lost interest and drove away when Mark said he wasn't.

"I hated you at that moment, but I kept reading your blog," he said. "I lightened up after you got hit in Florida. And there wasn't anything for me to hold onto once you broke my record. I really came around to root for you after that."

I was happy he had. Mark was an inspiration, and I knew our connection would last beyond December 31st. Looking to have a bit of fun

---

* Komito was pushed by two other competitors, Al Levantin and Greg Miller, and their competition inspired Mark Obmascik's 2004 book, *The Big Year*. That work was adapted into a motion picture of the same name in 2011.

† By the time of this book's publication, Hayward's record has been exceeded at least dozen times. Growing numbers of birders find more rarities each year, and social media broadcasts those discoveries to wider audiences faster than in the past.

with my new friend, I asked, "Would you have invited me to stay here if I came through here right after the dude mistook you for me?"

"Hell no!" Mark replied. "You'd be sleeping on the beach!"

That outcome thankfully avoided, we extended our conversation until one of Mark's friends, Malia, called with news from Half Moon Bay; a local expert, Alvaro Jaramillo, had found a Red-throated Pipit on the town beach. Superficially similar to a sparrow in appearance, the pipit nests in Scandinavia and Siberia and—like the Pacific Golden-Plover I'd chased and missed at several points—occasionally migrates to the West Coast instead of equivalent latitudes in Africa and Asia. I'd planned to spend the following day birding in Pescadero before continuing south, but backtracking north to pursue the pipit was a no-brainer. Half Moon Bay was only twenty miles away, and history suggested I wouldn't have a more convenient crack at the species.

Coordinating through Mark, Malia and I formulated a plan. She and her husband, Chris, would scour the beach at sunrise and call me if the bird was present. Vagrants often relocate overnight, so their assistance would ensure that I didn't chase a departed bird. A strategy in place, Mark and I hit the hay.

We rose early and cycled to the Pescadero beachfront while we awaited word from Malia. Pointing our spotting scopes at the horizon—"seawatching"—we scanned for pelagic birds, oceanic species that spend their non-nesting time wandering the world's oceans. Albatrosses are the classic pelagic example, but petrels, storm-petrels, shearwaters, and jaegers fall into the same bin. Pelagic birds often concentrate around nutrient upwellings like those along the edge of the continental shelf, and California is renowned for the pelagic boat trips that visit those offshore habitats.

Unable to join organized pelagic trips because they rely on petroleum-powered boats, I'd need to spot pelagic birds from land. That's an impossible task in most places because the sought species often stay tens or hundreds of miles offshore, but unique bottom topography along the Central Coast of California brings the continental shelf edge and associated deep water within sight of land. With patience and a good spotting scope, it's possible to observe a variety of pelagic species as they wing along the horizon.

West winds aid identification by pushing the birds closer to shore, and Mark helped me suss out Pink-footed and Black-vented Shearwaters as we scanned the Pacific on the morning of September 25th.

I answered my phone when it rang at nine a.m.

Malia reported, "The pipit is still here! We'll keep our eyes on it, and I'll text you updated directions if it moves. Now get going!"

I packed up my scope, bid Mark a temporary goodbye, and sprinted north along the rolling, wind-swept coast. I reached Half Moon Bay seventy-five minutes later and wheeled on the specified bluff. Peering onto the beach below my vantage, I saw Chris and Malia pointing toward a patch of strewn vegetation. I glimpsed movement, and binoculars confirmed the pipit for species #560. With a brown-olive back and streaky underparts, the bird is unremarkable to anyone but hardcore birders, especially without its summertime red throat.

I ditched my bike in the dunes and joined Malia and Chris on the sand. "Thanks for the help. You guys are the best!"

Malia was about to respond when British Chris beat her to it. "No worries, mate! Some bloke came along with a dog, but I shooed them away. I made sure he knew the bike-birding champion was on his way!" he said proudly.

He took my right hand between both of his and shook it vigorously. Wide-eyed and grinning, he seemed star-struck, like an English schoolboy meeting his favorite footballer. His pipit assist meant the world to both of us, and I was humbled to witness how others had adopted my project as their own.

Chatting with Malia and Chris reenergized me, and I continued to nearby Pillar Point Harbor for some midday birding before returning to Pescadero in the afternoon. I spent Friday, September 26th, birding around there and continued south to Santa Cruz on the 27th. Shacked-up with a pair of twenty-something birders in that hippie hotspot, I readied for my ride to Monterey. West winds would prevail through Wednesday, so my plan was to spend four days seawatching from that pelagic mecca.

My phone rang while I was blogging. It was Malia. "You're not going to believe this," she said. "There's a Pacific Golden-Plover on the same beach where we saw the pipit."

"Damn it! Two days too late!" I replied.

"How do you want to play this?" Malia asked.

Half Moon Bay was 50 miles behind me, so pursuing my nemesis would require 100 unanticipated miles once I returned to Santa Cruz. Allowing for extended searching if the bird proved difficult, I'd need to budget two full days for the chase. It would delay my Monterey arrival until Tuesday (versus Sunday) and slice my four-day pelagic window in half because east winds were likely to push the pelagic birds back out to sea after Wednesday. If I wanted to vanquish the problematic plover, then I'd need to surrender some of my oceanic opportunity.

I asked Malia, "Can you and Chris check on the bird tomorrow morning, like you did for the pipit?"

"Yeah, but we can't stick around because we have stuff to do. We'll be home late afternoon, so you're welcome to crash with us if you come back up here."

A plan came together. If Malia and Chris were unable to relocate the bird the following morning, then I would continue to Monterey as planned; if it stayed through the night, then I'd decide between chasing it or ignoring it. Chasing the bird would be a huge risk, one I wasn't sure I was willing to take by the time I passed out on Saturday night.

Malia called an hour after sunrise and reported the plover present. Encouraged, I checked the weather in Half Moon Bay, cloud to prevail until midafternoon. That was perfect; overcast skies would minimize beachgoers and the associated possibility they'd scare the plover away. I gobbled my breakfast and started north.

The first ninety minutes of that three-and-a-half-hour haul unfolded as anticipated; gray skies sheltered me from the sun as I undulated between low drainages and high bluffs. Unfortunately, the sun began to beat the clouds back in the ensuing hour, and my plover plans imploded when I arrived at a sun-drenched beach overrun with people. A token scan revealed only gulls, and I sank defeatedly onto the dunes while I contemplated my options.

Returning to Santa Cruz that afternoon would allow me to reclaim one of the two days I'd surrendered, but I decided to stay in Half Moon Bay because I thought a night with Chris and Malia would be fun. I was exhausted after the morning sprint, and there was a chance the plover would reappear when people departed the beach at the end of the day. I'd seen a nice mix of birds in Pillar Point Harbor post-pipit, so I headed that direction, thinking the plover might have sheltered there.

Reaching the harbor on a falling tide, I found an exposed mudflat at the base of one of the jetties. Beyond the expected Willets, Whimbrels, and Sanderlings, I glimpsed a butterscotch-colored blob as it darted behind a rock.

I scampered through tidal puddles for a better look. As I closed the distance to fifty feet, the bird revealed itself. Brown-and-tan with golden streaks across the face, crown, and back, the plover sauntered onto the mudflat and began feeding.

Yes! Yes! Yes! I'm in your head, bird! I own you! I'm the plover daddy!

I felt like I was candy-flipping—simultaneously on ecstasy and acid—but I got a hold of myself before I scared the bird away. When it walked to within twenty feet of me, I flopped onto my belly for eye-level photos. That I was soaked from shoulders to toes didn't matter; the intersection was miraculous given the disappointment I'd suffered an hour earlier. Returning to my bike after spending an hour with the bird, I knew Pacific Golden-Plover (#563) would stand alongside White-tailed Ptarmigan and Greater Sage-Grouse as one of the year's greatest victories.

I spent a celebratory evening with Malia and Chris and pounded out ninety miles to Salinas via Santa Cruz the next day. I reached Monterey and adjacent Pacific Grove midday on Tuesday, September 30th and spent that afternoon seawatching from Point Pinos. The west winds delivered a single Pomarine Jaeger (#564) among dozens of Parasitic Jaegers, and I was able to pull two Buller's Shearwaters (#565) from the mobs of birds. Nowhere else on the continent can such pelagic bounty be observed from the comfort of a cliffside bench. It was magical and restorative.

West winds on the thirty-first offered similar spectacle but failed to deliver Red Phalarope (not to be confused with Red-necked Phalarope from Salt Lake). At worst, the plover rendered the phalarope a wash; at best, I'd gained the plover over a bird I would have missed even with additional pelagic opportunity. Accounting aside, the plover was its own triumph.

The ocean activity shut down on Thursday, when the wind switched to the east, but I stuck around Monterey for three more days to deal with California Thrasher (#566), Yellow-billed Magpie (#567), and Lawrence's Goldfinch (#568). A bit of long-shot pelagic birding on those afternoons yielded nothing new, and I vacated Monterey on October 5th. Powering south through Carmel-by-the-Sea, I was looking forward to the Big Sur rollercoaster.

TWENTY-SIX

Straight into Compton

The Pacific Coast Highway clung to the oceanside cliffs, and I shifted into my lowest gear as the narrow thoroughfare snaked skyward. I'd dropped into and climbed out of several drainages through the morning, but this latest was the largest, my legs faltering as I forced my transport uphill. Reaching the apex, the view exceeded my imagination; beaming sun cast the Pacific in shimmering sapphire, kelp swayed at the behest of emerald rollers, and breakers kissed the plunging precipice in a tumult of turquoise. With sea lion barks urging me onward, I surrendered to gravity, that familiar "frenemy" whisking me down the backside of the promontory.

Rushing air silenced the pinniped echoes, and I leaned into a sweeping bend to conserve momentum. I floated over a short rise and dropped onto another downhill section, threat blending with thrill as I rocketed along the cliff face. The scenery made it difficult to keep my eyes on the road, and I nearly became fish food on two distracted occasions. Reaching the outpost of Big Sur, I initiated my search for California Condor.

Condors are enormous vultures; the California variety measures four feet long and ten feet across the wings. The species historically ranged through the southwestern United States and along the California coast, but the icon experienced a near-terminal population collapse during the second half of the twentieth century at the combined hands of poaching, power line collisions, and lead poisoning, the last suffered after ingesting shotgun pellets lodged in carrion. The long-lived scavengers reproduce very slowly, so their population couldn't withstand such heavy and unnatural losses. By 1987, the remaining twenty-two California Condors were in captivity.

From that nadir, the grim tale turns around. Breeding programs stemmed the decline, and captive-raised birds were returned to the wild in 1991. Extended management guided the condor's comeback, and as of 2014, the population had swelled to 400-some birds. The wild half of them sport color-coded wing-tags and radio transmitters, which relay positional data to researchers. Many free-ranging pairs breed without assistance, and the future of the species looked far brighter at my Big Sur arrival than it did in the late 1980s.

I intersected an example soaring over the coastal cliffs, the gargantuan bird dwarfing nearby Turkey Vultures. Keeping twenty-plus pounds aloft would be an energetic challenge without midday thermals, but the condor's massive wings allowed it to gain hundreds of feet on those rising columns of hot air. In awe, I watched the scavenger spiral toward the heavens. The incomparable species not ready to depart the land of the living, I was encouraged the California Condor would continue to amaze future generations.

I traversed the ruggedly beautiful southern section of Big Sur Coast the next day. Aesthetic considerations suggested I stick to the Pacific Coast Highway all the way to Los Angeles, but I needed to cut inland to look for LeConte's Thrasher, a mockingbird-relative that ranges through southern California, southern Nevada, and western Arizona. The gray bird sports a hooked beak and long tail and suggests a miniature roadrunner when it scampers across the desert floor.

My LeConte's quest began when I tacked east from Cambria, overcame the Santa Lucia Mountains, and dropped into Paso Robles on October 7th. The climate of that municipality is arid enough to produce wine, almonds, and olives, but the thrasher prefers habitat that is drier still, the scorched-earth sort found farther inland. Hydrating ahead of my ninety-five-mile ride to Taft, I readied for sweltering conditions.

I departed Paso Robles at sunrise, joined Highway 46 east, and covered fifty-three miles to Blackwells Corner. Turning south on

Highway 33 after eating lunch at that unremarkable intersection, I wasn't prepared for the nightmare that lay ahead. Clear skies allowed midday sun to beat down with unfiltered vengeance, and dehydration gripped my body as temperatures approached 100 degrees. The two-lane road was cracked and cratered, and what parts of the skinny shoulder hadn't crumbled were buried beneath drifting sand. With industrial litter strewn everywhere, my ride devolved into a worrisome weave around pipes, sheet metal, and other debris.

Bad as the road was, an unyielding parade of eighteen-wheeled tankers was worse. Unwilling to accommodate a cyclist, impatient drivers left just inches between their rigs and my body. Twenty seconds after one truck passed me, another came roaring toward me, the respective pressure waves sucking me into the traffic lane and shoving me into the desert. It was terrifying, but I had only myself to blame. Highway 33 wasn't built with cyclists in mind because no one of sound judgment had reason to ride into the superheated wasteland. Taking a rock off the helmet as another big rig overtook me, the sound of the strike reverberating like a gunshot, I felt like my LeConte's quest was a kamikaze cause.

The reason for the abundance of tankers became clear as I wheeled into the Midway-Sunset Oil Field, the largest petroleum extraction facility in California and the third largest in the United States. Conduits and collection tanks spread out as far as I could see, and thousands of derricks dipped and rose like metal mosquitos sucking the blood out of the planet. The heat was suffocating, and the presence of industrial candles was insulting; employed to burn off natural gas released as a petroleum by-product, they heated the landscape and increased the petrochemical stench to revolting levels. The scene was a nauseating example of the damage humans have inflicted on the planet.

I lost probably five pounds between sweat and fret, but focus and luck delivered me to Taft unscathed. I visited an ice cream shop, found a motel, and showered off petroleum by-products. Drifting off after blogging, I was proud my Big Year wouldn't sanction the environmental disaster I witnessed. Despite the day's dangers and struggles, the bicycle made more sense in the Midway-Sunset than it had anywhere else.

I rose early, ahead of anticipated heat, and explored Petroleum Club Road on the south side of town. The derricks along that stretch were rusted and decommissioned, and I listened for the thrasher's diagnostic call as I rolled through the dusty habitat. Prepared for an all-morning search, I was encouraged when I heard a two-note whistle after an hour. The bird dominated our ensuing game of cat-and-mouse, but I eventually enjoyed views of it and a companion as they sprinted between clumps of desiccated vegetation.

How LeConte's Thrashers thrive in such superheated surrounds is incredible. They've been observed drinking from natural water sources only a handful of times, and it's thought they extract all the water they need from their insect prey. At the opposite end of the topographical and temperature spectrum from the White-tailed Ptarmigan, LeConte's Thrasher rivals that alpine specialist for invincibility. More fragile than both, I retired to my motel; with mid-morning temperatures soaring and my legs sore after yesterday's Midway-Sunset gauntlet, the 100-mile ride to Ventura would have to wait.

That coastal return began before sunrise the following morning, October 10th, when I departed Taft on a safer stretch of Highway 33. The temperature and pitch increased across the ensuing forty miles, but I overcame oppressive heat and 4,900 feet of elevation gain to cross the Topatopa Mountains midday. From there, it was a relaxing downhill ride through Ojai and onto Ventura, where I rested. I then used a string of heavy-birding, low-mileage days to reach Santa Monica on October 14th.

I would have skirted Los Angeles on the Pacific Coast Highway, but I needed to venture into the sprawl to find Spotted Dove, an Asian species that was introduced to the city in the early twentieth century. The non-native thrived in neighborhoods and parks, and it radiated into Ventura, San Bernadino, Orange, and San Diego Counties. By the 1960s, the Spotted Dove was presumed to be a permanent addition to the region's avifauna. As such, the American Birding Association (ABA) decided it could be treated (meaning included in the count) the same as native species for listing and competitive purposes. Importantly, not every non-native is countable; a non-native must establish a breeding population, sustain itself for at least fifteen years, and be believed to be a permanent presence before the ABA considers it for countability. There is endless debate about which non-natives should count and which shouldn't, and some species are better established after ten years than others are after fifty. Most birders welcome the inclusion of non-natives because they push lists higher, but a purist faction reviles them because of the confusion and contradiction that their inclusion sows.

The Spotted Dove exemplifies that fallibility. Though the species flourished in Southern California through the 1970s, the bird suffered precipitous and unexplained declines in subsequent decades.* By my 2014 arrival, there were only scattered reports from the greater Los Angeles area, those representing hangers-on as the population finalized its apparent collapse.† There was, however, a handful of recent sightings in the city's South Central region, so I steered toward those neighborhoods from Santa Monica.

I left the beachfront at Venice, battled traffic through Culver City, wheeled through downtown, and reached Huntington Park after lunch. Slow-cruising those residential streets for the ensuing five hours, I found a single Spotted Dove sunning itself on a rooftop for species #573.

The dove's history and countability aside, my search for it offered an unexpected window to South Central. I was only fourteen years old when four Los Angeles police officers were acquitted of beating Rodney King, but I recalled the images of the ensuing protests as I explored the same neighborhoods where buildings burned in the spring of 1992. Outside of that brief window, most of what I knew about South Central

---

* Several hypotheses have been advanced to explain the Spotted Dove collapse—more Cooper's Hawks, an unidentified toxin, competition from the later-introduced and more aggressive Eurasian Collared-Dove—but there wasn't conservation impetus to understand the decline when the native population in Asia was thriving.

† As of 2014, a healthier Spotted Dove population persisted on Santa Catalina Island—twenty-two miles offshore—but I couldn't reach that destination without a powered boat.

was informed by the rap music that emanated from the area and the surrounding communities. I listened to N.W.A., The Pharcyde, Snoop Dogg, Tupac Shakur, and Ice-T through high school and college, and their rhymes suggested they experienced a very different America than I did. Voices for the underserved and neglected, they painted bleak pictures of the neighborhoods where I would seek the Spotted Dove two decades later.

Despite initial reservation and apprehension, I found South Central welcoming and safe. Streets featured pedestrians of various complexions speaking different languages, and no one seemed to mind the out-of-town cyclist with binoculars slung across his chest. South Central lagged behind other areas in economic development—abandoned buildings and vacant lots more common than in other areas—but the difference between my dated preconception and the contemporary reality was encouraging. Whether it was rolling through small towns in North Carolina or pedaling down Compton Avenue in South Central, the bicycle opened windows that skyways and freeways would have denied.

A two-night pause in Whittier lent recovery, and I rolled into Orange County at Huntington Beach before continuing to Newport Beach for the night of October 16th. Though the different municipalities between South Central and Newport displayed socioeconomic and ethnic variation, the character of each was overpowered by the collective sprawl; single-family homes and strip malls consumed fifty miles of coast and nearly double that distance inland. I marveled at the inefficiencies the decentralization perpetuated, none more glaring than the freeways.

Constructed in the mid-twentieth century as a short-sighted alternative to commuter trains—some suggest at the bidding of the automotive, petroleum, and tire industries—Southern California's freeways didn't scale as the region's population exploded. Traffic jams are an interminable headache, and engrained automotive dependence guarantees the problem will persist as long as the sprawl extends its gangrenous march.

My experience in Midway-Sunset exposed the supply side of the petroleum equation, but my view to the demand side was equally offensive; tens of thousands of vehicles clogged twelve-lane freeways as I navigated the cityscape. I hoped advances in renewable energy sources would lessen our dependence on petroleum, but I suspected weaning Southern Californians from their automobiles would be more difficult. Pedaling past stalled freeway traffic, I renewed the satisfaction I experienced after surviving Midway-Sunset.

Looking south from Newport Beach, I set my sights on Yellow-green Vireo. The neotropical vagrant had been spotted in San Diego on October 11th and—despite my prediction from Ventura—remained at Fort Rosecrans National Cemetery since its discovery, a positive report on the afternoon of October 16th buoying my hopes as I pedaled out of Newport on the morning of the 17th. The cemetery closed at four thirty p.m., so I needed to cover the eighty-seven intervening miles with enough time to look for the small and inconspicuous bird.

Following the beachfront, I sprinted through San Clemente, cranked through Carlsbad, and limped through La Jolla, my legs exhausted as I rounded Mission Bay and gained elevation onto Point Loma. I reached the cemetery at two p.m., dumped my bike, and began my search. I found the ficus trees the vireo had frequented, but strong wind made viewing difficult. My strategy for detecting arboreal songbirds relied on detecting movement, so spotting the sparrow-sized bird amidst swaying branches and shaking leaves seemed an impossible task. Scrutinizing three trees for five minutes each, I observed zero birds. A fourth ficus held only a single Yellow-rumped Warbler.

Wait—what was that?

A small bird departed the fourth tree and flew to a fifth at the cemetery boundary. I hustled in that direction and stared into the branches. Five minutes elapsed without sign of the escapee; the next ten offered nothing but a sore neck. Finally, after fifteen minutes of squinting, a chartreuse form resolved at the top of the tree, my hands shaking as much as the branches as I photographed the foreign visitor.

While species #577 was reason for celebration, I was particularly proud about reaching the four corners of the lower 48 states under my own power. Many Americans won't visit New England, Florida, Washington State, and Southern California in their lifetimes, so cycling between those distant destinations in a ten-month span felt like a monumental accomplishment. Departing the cemetery, I soaked in elevated views of San Diego from Cabrillo National Monument before continuing to my place of lodging for the night.

I spent the following week bouncing between La Jolla, Mission Bay, and Imperial Beach, where extended bouts of seawatching revealed Brown Booby, Cassin's Auklet, and Ancient Murrelet and brought me to within twenty species of my 600-species goal.

Looking east from San Diego, I faced 2,000 miles of riding to reach southern Texas. New birds would be few and far between as I retraced my tracks across southern Arizona, southern New Mexico, and western Texas, but my goal was within my grasp if I reached the Lower Rio Grande Valley by mid-December. With 14,100 miles amassed by the time I rolled out of San Diego on October 24th, I hoped my aching legs could manage the borderlands challenge a second time.

TWENTY-SEVEN

Desert Oasis

Among the twenty-some species of gulls that occur in North America, the Yellow-footed variety is visually unremarkable. Its dandelion feet and legs aren't diagnostic in isolation, and its dark gray back can suggest other *Laridae* family members. Only when those features are coupled with a stout yellow bill in the correct geography should the Yellow-footed label be applied. Generally confined to the Gulf of California in northwestern Mexico, the bird reaches into the United States at—and only at—the Salton Sea in Southern California, a bit of trivia making this inland body a must-visit for all Big Year birders. My plan was to visit the sea as I vacated Southern California and continued toward Arizona, New Mexico, and Texas.

I departed Jacumba Hot Springs on Interstate 8 and descended into the Imperial Valley, a vast agricultural swath in otherwise surrounding desert. Following dirt roads north through that cultivated expanse, I steered toward the southeastern shore of the sea, historically the best birding.

The Salton Sea is an unlikely intersection of natural forces, demographics, agriculture, engineering, and politics. Situated in the Salton Sink, 280 feet below sea level, the sea has cyclically filled and emptied depending on the variable path of the lower Colorado River. Though the contemporary flow empties into the Gulf of California, it has periodically shifted north and filled the Salton Sink to create an inland sea. The last natural filling cycling occurred in the seventeenth and eighteenth centuries, but the river's renewed southern course coupled with evaporation left the sea dry by the middle of the nineteenth century.

At that same time, the California gold rush of 1849 spurred thousands of fortune-seekers ("forty-niners") west. Many migrated through the Imperial Valley, and some folded their dreams of precious metals when they recognized the rich sediments that the meandering Colorado River had deposited along its former course. The region receives less than three inches of annual rainfall, so development companies constructed a network of canals to siphon water from the river. A farming industry took hold, and the Imperial Valley emerged as a stronghold of American agriculture.

Massive snowmelt in the Rockies overwhelmed the silted canals in 1905, and engineers initiated a project to relieve stress on the overflowing dikes. Their plan was a massive failure—one rivaling my first day in the biking shoes—and a catastrophic breach diverted the river's entire flow into the Salton Sink for the ensuing two years. Resorts sprung up around the impromptu oasis, and it was a popular vacation destination until the 1970s, when the cycle of agricultural runoff and evaporation rendered the sea too salty and polluted for recreation. The resorts were abandoned, and the surrounding desert reclaimed the decaying encampments.

Birds have been the primary attraction to the Salton Sea since the human exodus. A hearty tilapia population supported tens of thousands of pelicans and cormorants through the end of the twentieth century, and hundreds of thousands of migratory birds spend the winter at and around the sea because of its position along the Pacific Flyway, a major migration route. Coupled with the prized Yellow-footed Gull, that biomass makes the Salton Sea a premier birding destination between October and April, the only months when temperatures are bearable.

I intersected the sea's southeastern shore midday on October 25th. The body's signature stench, a festering cocktail of agricultural waste and decaying fish, welcomed me to its shores. Tilapia die-offs have become common as the sea contracts because evaporation increases salt concentrations and reduces dissolved oxygen. Tilapia can survive osmotic and pollution challenges but inevitably succumb to hypoxia, thousands of their skeletal remains crunching underfoot as I wandered the shoreline.*

Moving northeast toward Obsidian Butte, an elevated mound that affords sweeping views of the southeastern shoreline, I found thousands of birds. Heat waves blurred Herring, California, Ring-billed, and Laughing Gulls into unidentifiable mobs, but I continued scanning with my scope because the darker-backed Yellow-footed would contrast against those paler hordes.

I spotted a sooty blob a half-mile away and walked down the shore-line for a closer view, soft mud stopping me when I'd closed the distance to 150 yards. The bird looked promising, but I needed to see the legs to eliminate the pink-legged Western Gull, a species I'd seen daily for the previous two months. Unfortunately, a pair of husky Herring Gulls obscured my view of those appendages, so I waited for the blockers to move. Just when I thought the obstructing birds had grown roots, they finally kicked their asses into gear, waddled out of the way, and allowed a clear view to the banana-legged bird behind them.

Yellow-footed Gull for #582! You're mine, you Salton Sea son-of-a-bitch!

My view was distant, but I hardly cared. The gull was the last bird I needed to find in California, and with it secured I could turn my attention east. My timely find suggested that I should forfeit my second Salton Sea day and continue immediately into Arizona, but I decided to hang around for a second day as planned. The sea's future uncertain, I wanted to enjoy it before the desert reclaimed it.

I overnighted in Brawley and retraced my tracks to the sea the following morning. The first half of that return was uneventful, but I was struck with sudden and painful intestinal distress as I approached Obsidian Butte. No bathroom for miles in any direction, I jammed on the brakes, scampered behind a boulder, and completed the purge with my helmeted head peering over the rock. Dread gripped me when I

---

* By the time of this publication, few—if any—fish remain in the Salton Sea, and tens of thousands of pelicans and cormorants have relocated in the absence of food. Dramatic flux is likely to continue as the sea continues its contraction.

discovered my cache of toilet paper was missing, and I rifled through the bag for a substitute. When none availed, I turned to my surroundings.

Can I get a paper bag, a burger wrapper, an old magazine, any god-damned thing?

Nothing. I hung my head. And there, in my earthen gaze, my salvation appeared. I unlaced my shoes and peeled off the underlying socks. Performing the deed, I discarded the socks and covered the scene with several rocks. I survived the day without further episode and returned to my motel late in the evening.

That second Brawley night, October 26th, I learned a Rufous-backed Robin had spent a second day in Twentynine Palms, a small town just north of Joshua Tree National Park. I'd been alerted to the Mexican vagrant the previous day, while I was viewing the Yellow-footed Gull, but I dismissed the possibility of a pursuit because Twentynine Palms was 150 riding miles northwest of Brawley. I was headed southeast, toward Arizona, and I thought it would be insane to invest three or four days into a wandering bird for which no behavioral pattern had been established. Putting it out of mind, I spent my second scheduled day at the Salton Sea. Minus explosive diarrhea, it was lovely.

However, my calculation changed when I learned the rare robin spent a second day frequenting the same fruiting palms and adjacent pond. It was a perfect place for the vagrant to rest and recover, but the bird would need to stay through two more nights, the 26th and 27th, for me to have a crack at it midday on the 28th, the earliest I could reach Twentynine Palms. By the time I backtracked to Brawley, the 300-mile detour would be equivalent to riding from Boston to Philadelphia, St. Louis to Chicago, or Vancouver to Portland. I'd made equivalent deviations for bonus birds, but all side quests sought resident species represented by hundreds or thousands of individuals; if the lone and transient Rufous-backed Robin abandoned the palm grove before I arrived, then my efforts would be wasted. Powering north out of Brawley on the morning of the 27th, bike-birding felt exactly like drinking; I couldn't stop myself from binging.

I followed Highway 111 through Calipatria and along the eastern shore of the sea to reach Bombay Beach. Another two hours of cranking in ninety-degree heat put the sea behind me, and positive reports of the robin through the afternoon buoyed my hopes as I navigated Coachella and Indio. By the time I reached North Palms Springs, I'd covered 111 miles. If the robin was obliged to stay through that night, then I'd have a chance at it the following morning.

My pursuit resumed on the 28th before sunrise with a 22-mile, 3,000-foot ascent on Highway 62. From the summit at Yucca Valley, it was an equivalent downhill distance to Twentynine Palms. A half-dozen birders hadn't observed the robin by my ten a.m. arrival, so I settled into the collaborative search.

Fifteen minutes later, my heart quickened when a bird of the appropriate size and shape flew into an overhead palm.

Was that the Rufous-backed? Or was it just an American Robin?

My glimpse was too brief to eliminate the more common relative, so I focused on the tangled fronds. My bulging eyes were burning after two unblinking minutes, but I detected a slight rustle in the vicinity where the candidate had vanished. Another twitch and a bird emerged, the Rufous-backed stealing my breath as it gobbled palm fruit. I shouted to the others, "It's over here!"

They joined me, and we spent the next half-hour trailing the bird around the palm grove, cameras clicking throughout. Some petroleum-powered Big Year birders are willing to throw thousands of dollars at individual species toward the end of their campaigns, but I doubted any of them had invested more in the pursuit of a single bird than I had in the robin. It was the year's best feeling: a natural high no amount of alcohol or drugs could replicate.

Ecstasy yielding to pragmatism, I departed Twentynine Palms, returned to Yucca Valley, descended to North Palm Springs, and continued to Indio by nightfall. Coupled with the 111 miles I made on the first day, 114 more on the second left me 79 miles to reclaim Brawley on the third. Rolling into that destination exhausted, I sought refuge at the same motel I'd departed three days prior. I'd pedaled 304 miles and not gained an inch toward Texas. I could, however, claim the rare robin, and that was enough at that moment.

A large pepperoni pizza in my clutches, I settled into Game 7 of the World Series between the Kansas City Royals and the San Francisco

Giants. It was October 29th, 2014, exactly six years since I met Sonia at Professor Thom's.

———

Sonia left me on January 9th, 2010, and I returned to Alcoholics Anonymous ten days later, three years since I'd quit the program to end my first sober stint of thirty-eight days. Establishing a routine to stay busy and avoid temptation, I woke at six a.m., walked to my lab in the dark, spent the entire day doing experiments, attended my nightly AA meeting, and returned home. Weekend birding outings to Central Park and Jamaica Bay offered occasional distraction, but my DJing interest faded because the music was a reminder of the disinhibited euphoria I'd never again experience.

My rituals were insulating and protective, but I broke from them to travel to Hawaii on February 9th, just twenty-two days into my renewed sobriety. Originally planned as a mid-winter escape with Sonia, the trip devolved into independent vacations after our split. Neither of us wanted to forfeit the plane ticket or experience, so—working entirely through email in the week following our split—we retooled the trip to create itineraries without the possibility of intersection.

I arrived in Kaua'i and spent five days exploring the Nāpali Coast, Kīlauea Point, and Alakai Swamp, those natural areas a welcome break from New York's concrete confines. Nightly refusals of alcohol and marijuana in campgrounds strengthened my confidence, and I was encouraged to know my sobriety could survive outside of my Big Apple routine.

I flew from Kaua'i to the Big Island on Valentine's Day. Watching airport couples canoodle on that occasion, I thought about Sonia. I'd been alone for thirty-five days, and few of my sober twenty-seven had passed without regretting the pain my behavior had caused her. Trapped in my tent by torrential rain that evening, a combination of guilt, curiosity, and boredom spurred me to call her. My advance wasn't invited, but I figured she could ignore the call if she didn't want to talk.

From O'ahu, she answered. "Hello?" she asked incredulously.

It was the first time I'd heard her voice since she fled my apartment. Caught between hope and regret, I was unable to speak.

"Hello? Are you there?" she begged.

I replied diffidently, "Yes, I'm here."

"Is this an emergency?" she asked. "Is something wrong with your trip?"

"No. I just wanted to hear your voice," I said. "I miss you."

She replied, "What am I supposed to do with this, Dorian? I'm exactly where I was a month ago. Nothing has changed."

Her prompt created the perfect opportunity to advertise my sobriety, but I resisted. I'd undertaken that journey independently of her, and I didn't want to add to the confusion my call had already caused.

"OK. I won't bother you again, I promise," I said. "Please take care of yourself, Sonia."

"You too," she said before hanging up.

I thought about our exchange for my final four days on the Big Island, but I put Sonia out of mind after I returned to New York. I realized I couldn't grow my sobriety if I was fixated on the past.

I resumed my routine but gravitated to my laboratory in the ensuing month. My daily work hours swelled from ten to fourteen, and I regularly crashed on the department couch rather than walk the mile home. Apart from occasional birding outings, weekends were structured similarly; it wasn't long before I was working eight to ten hours on Saturdays and Sundays, too.

My commitment to AA slipped as lab work consumed me. I had only eight months to complete a massive panel of experiments and write and defend my thesis before moving to Boston to start my post-doctoral fellowship, and nightly meetings became a resented disruption as my focus constricted on science. AA was a wonderful community and an initial source of structure and accountability, but it didn't evolve night-to-night and ultimately proved too passive for me. My addictive tendencies needed something to chew on in the absence of alcohol and drugs, and immersing myself in cell polarity felt the most natural way to avoid relapse. By mid-March, I'd attended my last AA meeting.

At about that same time, a month after our Valentine's exchange, Sonia sent me a text message asking how I was doing. Her outreach floored me, and I replied to let her know I was surviving. That wasn't enough for her, and a clumsy volley of follow-ups culminated with an invitation to meet at an East Village coffee shop the following after-noon. Curious and confused, I accepted, the possibility of a pregnancy making sleep impossible.

I arrived first and claimed a table in the back of the bohemian beanery, an ironic rendezvous point as neither of us drank coffee. Sonia showed five minutes later, her brown eyes sympathetic as she glided across the room. I offered a hug; she tentatively accepted. Against our emotional history, neither of us knew where to begin.

We tiptoed through pleasantries, but Sonia initiated a deliberate tack. She explained, "I wanted to see you because I've been attending Al-Anon for the last few weeks."

I was surprised because the group is a support network for the friends and family of alcoholics. She'd unburdened herself of me, and I didn't think she was friend or family to another alcoholic. "Really? Why?" I asked, my curiosity genuine.

"Because I was angry with you and couldn't manage those feelings," she said. "I thought there was something wrong with me, but the group helped me realize the problem isn't mine; it's yours. You're an alcoholic."

She spoke the truth, but it stung to hear it from someone else. I steadied myself and replied, "You're right, but I don't understand why you wanted to see me."

"Because I needed to confront you to let my anger and frustration go," she said. "I'm hoping this will bring me closure."

I understood and had no reason to deny her catharsis. "I know how badly I treated you," I said softly, "so I'll do anything I can to help you feel better."

"I will, in time," she continued. "And I hope you understand your drinking will eventually hurt you more than it did me. You have so much to offer if you can get sober."

She'd offered the perfect opportunity to advertise my sobriety, but I paused. Our encounter was awkward and emotional, and I wasn't sure I wanted to include her in a journey I'd undertaken for myself. I faced an immeasurable amount of work, sobriety a lifelong battle fought over individual days, but I figured she'd want to know I was confronting my condition.

"I haven't had a drink in two months," I said. "I didn't tell you over the phone, while we were in Hawaii, because it was my own thing. But it feels different with you sitting here."

She reached across the table and squeezed my hand. "I'm so happy to hear this, Dorian," she said. "I was scared you might have gone the other way, but I had to confront you either way."

I replied, "I get it. And I'm happy you did. At least you know I'm getting better."

She was crying. "You can do this," she said. "And I want to help. You can call me if you feel yourself slipping."

I didn't expect her to embrace my nascent sobriety given the pain my alcoholism had caused her, but her reflexive desire to help was one of her most endearing qualities.

"Thanks, Sonia. That means a lot," I said. "Sobriety is really isolating when everyone you know drinks to some extent. It's nice to know I have your support."

I promised to keep her updated on my progress, and we parted after an extended embrace. I didn't imagine resurrecting our relationship, but it was wonderful to know she wanted to be included in my sobriety. Initiated as a personal project, my recovery had sudden and unexpected support.

———

Powering south out of Brawley the following morning, I thought about how my life would have been different had Sonia not contacted me. It's possible my sobriety would have survived without her, but it was great to have her in my corner. I couldn't have foreseen it from that newly sober vantage, but our East Village interaction would go a long way to setting my bicycle Big Year course four years later. That transcontinental odyssey not yet complete, I departed Brawley and joined Interstate 8 toward Arizona, the Big Year clock ticking as I pedaled toward Texas.

TWENTY-EIGHT

Mission Accomplished

The misty air was cool, and my breathing reverberated through the predawn quiescence as I gained elevation into Saguaro National Park, Arizona. As the sun climbed off the horizon, golden rays dissipated the vaporous shroud and illuminated a beautiful array of Sonoran Desert flora. Among scrubby mesquites, bushy acacias, nimble ocotillos, succulent agaves, and paddled prickly pears, towering saguaros were the most prominent. Their barrel chests and outstretched arms suggested cheering spectators, and an imagined gallery roar—the motivational sort that propels marathon runners toward the finish line—lent encouragement as I labored through the landscape on November 4th, five days after departing the Salton Sea.

My return to Tucson closed an 8,430-mile loop initiated from the same city on May 31st. In the intervening five months, I'd conquered the Rockies, navigated the Northwest, traversed the Pacific Coast, and returned to the desert, a circuit pushing my bird list to 583 species and my mileage beyond 15,000. I'd cultivated incredible cardiovascular fitness—my resting heart rate was less than 50 beats per minute—but the rest of me begged for extended recovery. My leg muscles had the fortitude of marshmallows, my backside felt like it had endured ten semesters of fraternal paddling, and my joints ached from the pounding imparted by cracked and cratered roads. Facing two additional months of riding, I wasn't sure if my battered body could absorb additional abuse.

My Tucson return also signaled a temporary end to novelty because I'd rejoin Interstate 10 and retrace my tracks to Central Texas over 650 familiar miles. I added 100 new species along that trajectory in spring, but I expected only two new birds on my backtrack: Baird's Sparrow and Sprague's Pipit, both of which nest in the Upper Midwest and winter in the Southwest. With few new birds, my eastbound return would be a slog.

Against that certainty, the Sinaloa Wren I added in Tubac (south of Tucson) was a welcome surprise. The brown skulker is endemic to western Mexico but makes rare cameos in Arizona, the Tubac bird just the fifth recorded in the US. The intersection was anticipated because the vagrant had occupied the same suburban thicket for six weeks, online reports keeping me updated as I approached, but it was an unexpected bonus nonetheless. Spirits boosted by species #584, I rejoined I-10 at Benson and continued east into New Mexico.

Tailwinds sped me through Lordsburg, Deming, and Las Cruces, but my fortunes sank when a reinvigorated polar vortex thrust subfreezing temperatures onto West Texas. The sixty-mile ride from El Paso to Fort Hancock was miserable without insulated clothing. The icy wind froze my fingers to my handlebars, and I took periodic refuge in gas stations to renew blood flow to my extremities. Conditions improved as I reached Van Horn and Fort Stockton in subsequent days but deteriorated again as I approached Ozona; the 110-mile ride to that destination required eleven grueling hours, a murderous northeastern headwind blowing at 20 miles per hour through the afternoon. Staggering into Sonora on November 17th, I'd covered 400 miles in the five days since departing El Paso.

I broke from I-10 at Sonora, overnighted in Rocksprings, and rolled out of Hill Country on the 19th, the juniper-oak flora yielding to mesquite-dominated scrub as I dropped off the southern edge of the Edwards Plateau and sped toward Uvalde. A sighting of Great Kiskadee—a robust neotropical flycatcher with a yellow belly, rufous-brown back, and black-and-white-striped head—suggested progress, and Green Jay, Long-billed Thrasher, and Olive Sparrow solidified my South Texas arrival. Reaching the lower throws of the Rio Grande Valley at Mission on November 26th, I claimed 598 species.

I continued east on Thanksgiving morning, anticipation growing as I approached Estero Llaño Grande State Park in Weslaco. Upon arrival, I was greeted by three park staff and Bill Sain, a sixty-something birder and faithful blog reader with whom I'd been communicating. The quartet shepherded me onto a low boardwalk. Bill pointed at a ripple along the edge of the impoundment.

"You still need Least Grebe, right?" he asked.

I raised my binoculars as the small swimmer surfaced. "Not anymore," I bellowed. "And that's 600, since a flock of Green Parakeets flew over me on my ride here!" I let my binoculars dangle, clenched both fists, and continued, "It only took eleven months and 16,000 miles, but I did it!"

"Congrats, Dorian," Bill said. "It's an incredible achievement."

Javier, one of the rangers, chimed in, "And it's great you found number 600 at Estero. We're happy to be a part of history!"

"Thanks, guys," I replied. "I thought 580 was realistic when I started, but I decided to aim for 600 because it sounds better, not because I thought I could do it. Now that it's happened, it doesn't seem real."

I stepped aside and took a moment to reflect. Born out of scientific frustration and personal confusion, my bike-birding intention was initially absurd; there was no precedent to orient my quest, and resigning life-long academic aspiration to pursue the most difficult and dangerous Big Year in history was a dramatic departure from the course I'd steered for the previous sixteen years.

Since taking that headlong leap, I'd explored twenty-eight states, experienced an amazing array of American landscapes, and interacted with an inspiring sample of people, each of whom left a stamp on my transcontinental travels. I'd overcome everything the open road had thrown at me, and I'd reclaimed the adolescent Big Year dreams that academia and alcohol had suffocated across two intervening decades.

Most important, I'd freed myself from external standards and expectations. Primacy granted me, the first person to undertake a nationwide bicycle Big Year, a reprieve from comparison because there weren't established routes or existing records to serve as references or benchmarks. I wasn't chasing anything but my own imagination, and I was free to define success however I wanted. On some days, that was finding a specific bird; on others, it was turning the pedals when every cell in my body wanted to quit. Each day presented new challenges, and I found unexpected pieces of myself with each hurdle I overcame.

Gazing over the pond, I realized 600 species wasn't the end of my journey; it was the beginning. Leaving academia was a gut-wrenching decision, but everything I'd experienced since that pivot indicated I'd be happier doing something else. I didn't know exactly *what* at that Estero moment, but I finally understood the future wasn't an ideal or static construct. It was an opportunity, my bike-birding adventure suggesting everything was impossible until I decided it wasn't.

Satisfied with my direction, I turned to my companions and asked, "What's next?"

Departing the boardwalk, my escort led me to a Common Pauraque, a nocturnal insectivore that reaches into the United States only in South Texas. Pauraques are eleven inches long, and their intricate brown plumage and bulbous heads suggest moths as much as birds. They disappear against leaf litter, and I would have overlooked the bewitching bird had the park staff not alerted me to it, its terrestrial roost barely three feet off the path.

Looping back to headquarters after intersecting Groove-billed Ani (a black cuckoo with a knobby beak) and Buff-bellied Hummingbird, our assembly received news of a curious bird in another part of the reserve; described as small and green with a curved beak, it warranted investigation. Accompanied by a pair of birders who had since arrived, we hustled to the referenced area and dispersed into the habitat. Our hour-long search yielded only expected species, and we gravitated to a shaded area as midday heat suffocated our enthusiasm.

Mary Gustafson, a long-time valley resident and local birding expert, interrupted our subsequent socializing. "Honeycreeper!" she shouted.

I wheeled around to see her pointing at a tree. My eyes seized on a small blob in the lowest branches. My binoculars jumped to my face, and a frantic spin of the focus wheel brought a green bird with a stumpy tail and curved bill into focus.

"What the hell?!" I exclaimed. I released my binoculars and reached for my camera, but the bird flew into the adjacent thicket before I completed the swap.

I turned to Tiffany, a local twenty-something who had joined our search. She was wide-eyed, as though she'd seen a ghost. "What on Earth was *that*?" she asked disbelievingly.

Our assembly parsed a more precise identification. When our combined recollections suggested a female or immature Red-legged Honeycreeper, a species that hadn't been observed in the United States previously, we were ecstatic—until we realized no one photographed the bird. Without corroborating evidence, our sensational sighting could be rejected by governing bodies, a fate akin to winning the lottery then losing the uncashed ticket.

The next hour was an emotional roller coaster as we tried to relocate the honeycreeper. Every small bird elicited excitement until eliminated as something else, but our attentions eventually converged on a promising candidate. The bird was descending a large tree, and anticipation built as it neared our terrestrial vantage. My camera ready when it flitted onto an exposed, eye-level branch, I squared the tiny target in the viewfinder and pressed the shutter. The bird fled in the next instant. My fingers trembled as I consulted my camera's digital display; it showed a Red-legged Honeycreeper in perfect focus.

Others had also managed pictures, and our assembly started hooting and hollering like we'd won the most thrilling Super Bowl in history. We were birdwatching champions, imagined confetti falling from the Texas heavens as quarterbacks, cheerleaders, and billionaire owners stood and applauded.

Breaking from delusion, we disseminated the news of our fantastic find through online channels. Within an hour, local birders had disregarded family, abandoned Thanksgiving feasts, and raced to join the honeycreeper hubbub; one conflicted man arrived with a half-eaten drumstick in his hand. Our entourage ballooned to twenty, and we broached the inevitable discussion of provenance—how a vagrant reached its discovery point—while we tried to relocate the bird for the latecomers.

To count toward standardized or competitive purposes like Big Years, vagrants must be believed to have reached their discovery points without assistance. A flightless Common Ostrich could not reach Iowa from Africa unless transported by humans, so a free-running example in Des Moines would be an obvious escapee and summarily discounted. Unfortunately, judgment is rarely so straightforward. Birds are routinely trapped and transported, and records committees assume the unenviable task of determining provenance from incomplete evidence. Waterfowl are particularly problematic. They are capable of prolonged overwater flight and subject to disorientation during migration, but the preponderance of ornamental collections—and the associated possibility of escape—subjects sightings of rare ducks and geese to intense inspection. Anomalous birds of prey, and particularly falcons because of their use in falconry, are subjected to similar scrutiny. No birder wants to chase a vagrant if it's likely to be invalidated after the fact, and records committees use photographs, written accounts, and decades of anecdotal experience to mete out judgments on a case-by-case basis.

I'd observed many unusual birds during my Big Year, but there was a historical pattern of vagrancy for each of those species. Sinaloa Wren, for example, was understood to stray to southeastern Arizona by the time I observed it, so adding it to my list was a formality. There wasn't, however, an equivalent precedent of vagrancy for Red-legged Honeycreeper, and that circumstance required special consideration before I could count the bird toward my Big Year total.

The Red-legged Honeycreeper ranges through Latin America and reaches north into Central Mexico, within 400 miles of the United States; it was easy to imagine a disoriented or adventurous individual wandering north to Texas. The stunning, cobalt-colored males are kept as pets under rare circumstances, but there was little chance anyone would catch and transport the olive female or immature individual we'd observed. The bird wasn't banded or marked, our photographs didn't reveal feather wear indicative of captivity, and the subject fled us as a pet bird might not. The available evidence suggested our honeycreeper was wild, and I could include it in my total if the relevant records committees agreed. Administrative backlog guaranteed the review process would take at least a year, so I put the species aside as "provisional" until we submitted documentation and received a ruling.

Though our expanded search party agreed with our preliminary provenance assessment, we were unable to relocate the bird for the latecomers.* Adrenaline had suppressed midday hunger, but my grumbling stomach would not be denied as evening approached. I thanked my companions for an unforgettable day and powered out of the reserve.

Racing the setting sun toward my arranged lodging, I couldn't believe how the day had unfolded. The delayed gratification of reaching the 600-species plateau was one sensation, and the spontaneous euphoria of the honeycreeper was another; I felt like I'd augmented a beer buzz by snorting a line of cocaine. Happy I'd found a substitute high, I continued to the Alamo Inn B&B, a birding-themed property boasting a gift shop, in-house birding guides, and an open kitchen, that last amenity a godsend because restaurants were closed for Thanksgiving. Tucking into a quadruple layer turkey sandwich, I had much to be thankful for.

—————

Sonia's pledge to support my sobriety lent motivation and accountability after our rendezvous in the East Village coffee shop. When I made it through the following week without drinking, I contacted her. She hadn't indicated her desired level of involvement, so I texted only the critical information, "Another sober week. 63 days and counting."

She replied, "That's great. Keep me posted."

I hadn't advertised my daily progress since abandoning AA, so her acknowledgment, however minimal, was encouraging. A similar text the following Monday received a similar reply, but I offered more in round three. I wrote, "77 days. Made a cameo at departmental happy hour without melting down, so that's good."

Sonia seized on that accomplishment and replied, "That's a big step. Keep it up. And feel free to reach out if you're wobbling. I'm here."

By mid-April, I'd been alcohol and drug free for three months. I sent Sonia a text with that news. She wrote back, "Those are words I never imagined from you! So happy for you!"

---

* The bird resurfaced for dozens of observers on Friday the 28th and Saturday 29th before disappearing for good.

The injection of humor, however subtle, was uplifting because it showed Sonia could finally laugh about the situation. While I was picturing her smile, my phone pinged again. Her follow-up read, "The weather is perfect, so I'm going to walk to work. You want to join me?"

My knees buckled. I backpedaled, bumped into the couch, and sat. Assuming I maintained my sobriety, I imagined interaction beyond text messages was possible, but I hadn't dwelled on a timetable because the next move was Sonia's to make. Unprepared for in-person interaction on that morning, I nonetheless seized the opportunity to see her, a short walk the perfect chance to get reacquainted.

She rang my buzzer twenty minutes later. I told her I'd be right out and locked up, perspiring palms making that routine task difficult. Exiting the vestibule and stepping outside, I saw her, the surrounding hustle and bustle irrelevant the moment our eyes met. I had no idea what to say, so I was relieved she spoke first.

Extending an arm, she offered a bag from Dunkin' Donuts. "I brought you something to celebrate three months," she said.

Her gesture was unexpected but on brand, Sonia a master at making others feel special.

"Thanks," I replied. Fingering the sugary offering, I continued, "It's a daily battle. The winter sucked, but I've done better recently, especially with spring migration starting."

"You're still birding?" she asked.

"Yeah. The interest is really coming back strong. Morning outings are my reward for not getting hammered the night before, ya know?"

I asked how work was going and what she'd been up to on the social front; she summarized the previous three months before inquiring about my research. Though mostly transactional, the conversation was steady as we pounded the pavement. Our commutes diverged after fifteen minutes, and we agreed to repeat the walk the following Monday.

When that walk went well, I invited Sonia to go birding in Central Park. She accepted, and we spent Saturday morning taking in views of warblers, vireos, and thrushes. Afterward, while returning to the subway, Sonia made an off-the-cuff comment about being thirty and feeling old.

"I forgot you had your birthday in January," I said.

There was an awkward pause. I helped plan the party but didn't attend because Sonia left me two weeks ahead of the event. My reference to that tumultuous time deflated her, her smile gone and her arms hanging heavy.

"You . . . you . . . you had other stuff going on. I didn't expect you to call," she said.

I felt awful that I wasn't around to celebrate her special day. "I'm sorry, Sonia. That was a crappy time for both of us," I said.

She looked at her feet and spoke, "Yeah, well. It's done. Let's not rehash it."

We reached and boarded the train in silence. Approaching the stop closest to my lab, where I'd planned to spend the remainder of the afternoon, I asked Sonia if she wanted to extend our string of Monday morning walks.

"I need a bit of space right now. Give me a few days."

I understood her hesitancy. Beneath three-and-a-half months of sobriety, I was still an alcoholic whose drinking had caused her repeated emotional anguish. No amount of time or therapy would change that, and Sonia would need to consider my underlying condition in any future dealing with me. Though the morning continued our measured rebuild, I knew no future association was guaranteed. I wished her a good afternoon, exited the train at 33rd Street, and walked to my lab.

We didn't coordinate our Monday commute, but we touched base in time to make Thursday work. Our rapport tentatively reestablished that morning, I told Sonia about a plan I'd hatched. "It's something silly. Does Sunday afternoon work?" I asked.

"Yes," she replied. "But please keep it simple. And in public. We're not ready to be alone."

I'd considered that angle and planned accordingly. When Sunday arrived, I swung by Sonia's apartment. She came outside. I told her about my plan.

"I feel bad about missing your birthday, so today is a do-over. And since you said you were feeling old last weekend, I've planned an afternoon to make you feel young."

Sonia didn't respond, so I reassured her. "It'll be fun, I promise. Our first stop is only a few blocks away."

We walked to S'MAC, a macaroni and cheese restaurant that serves creative twists on the classic. I explained, "A while back, you mentioned that your mom made you Kraft Mac & Cheese after school. I thought this place might bring back some childhood memories."

Sonia beamed. "I love this place. I've been here before but it never gets old—unlike me!"

We gorged on crusty, truffle-infused pasta before waddling to the subway. Exiting at 59th Street, we visited Dylan's Candy Bar, a two-story dose of diabetes that makes customers feel as if they're walking through the childhood board game Candy Land. Fudge in hand, we walked to the Central Park Zoo.

"What kid doesn't love going to the zoo?" I asked rhetorically.

"This is so much fun!" Sonia exclaimed. "I feel like I'm ten years old. Thank you for this."

That week, we walked to work together twice. When the weekend rolled around, we attended a street fair hand-in-hand. We ate, laughed, and danced, and an impromptu kiss solidified the progress we were making.

Though my sobriety hadn't wavered across four months, I knew renewed affections upped the ante because a misstep with alcohol or drugs would drag Sonia down alongside me. I'd pulled her into the emotional depths too many times, and I didn't know how either of us would handle similar challenge in the future.

It was then mid-May of 2010, and I'd be moving to Massachusetts to begin my postdoctoral fellowship at the end of the year. I could not turn down the opportunity that Josh Kaplan, my future advisor, was offering me at Massachusetts General Hospital. Though Sonia suggested she'd be open to the possibility of a move when I accepted the position in the fall of 2009, my subsequent behavior and her resulting anguish demanded that she reassess those plans. June rapidly approaching, neither of us knew how to forge a future before distance made that task more challenging.

Chasing my turkey sandwich with a piece of apple pie, I realized the similarity between that time and my bicycle Big Year, the calendar forcing my hand in each instance. I'd accomplished my goal of 600 species with thirty-four days to spare, but my addictive personality guaranteed only temporary satisfaction before it refocused on the remaining possibilities. Scouring the Rio Grande Valley for two additional weeks, I observed White-tailed Hawk, Hook-billed Kite, Aplomado Falcon, Ferruginous Pygmy-Owl, Red-crowned Parrot, and Tropical Parula to claim 610 species on December 12th. The following morning, I followed Highway 77 north to Kingsville, a distance of 107 miles, before tacking slightly east and covering another 83 miles to Rockport on the 14th. The angelic Whooping Crane added on the shore of Aransas Bay on the morning of the 15th, I readied for the aquatic phase of my journey.

TWENTY-NINE

Best Friends

The American Flamingo ranges through the Yucatán Peninsula, Caribbean islands, the northern coast of South America, and the Galapagos Islands. The elegant wader once graced the Gulf Coast and South Florida, but feather traders hunted that subpopulation to extinction in the nineteenth century. Though representatives are occasionally observed in their former US range, questions of provenance are usually damning given the abundance of ornamental captives in the region. Some sightings certainly represent natural vagrants, but it's virtually impossible to parse wild examples from escapees—except in the case of "HDNT."

American Flamingo HDNT was captured as a wild juvenile on the Yucatán in August of 2005, and researchers affixed him with a leg band emblazoned with those letters before releasing him for observation. His jewelry can be read from some distance with a spotting scope, and it unequivocally identified him when he appeared in Rockport, Texas, in November of that year, his translocation likely aided by Hurricane Rita as she swept across the Gulf of Mexico in late September. Even if storm blown, HDNT was a wild bird, so birders rushed to Rockport to view him.

Many vagrants vanish soon after their discoveries (like our Red-legged Honeycreeper), but HDNT remained around Rockport for months. More remarkably, he was joined by a Greater Flamingo early in 2006. That much paler species ranges through Africa and Asia, so

Greater Flamingos in the United States are universally discounted as escapees because the Old World fixture couldn't reach the Americas unless transported by humans.

Such was the case with "492," also identified by his leg band. He and thirty-nine other Greater Flamingos were shipped from Tanzania to Kansas in 2004. There was a miscommunication during the exchange, and zookeepers failed to clip the birds' wings, an oversight that bit them in the ass when 492 and an unspecified accomplice escaped in June of 2005. The accomplice wasn't seen again and presumed dead, but 492 was spotted in Wisconsin that fall before disappearing ahead of winter. His survival and whereabouts were a mystery until early 2006, when he surfaced in Rockport, in the improbable company of HDNT. Their long-shot intersection was a heartwarming coincidence, and the best friends spent the following eight years exploring the Texas and Louisiana coasts together.* Only HDNT counted toward Big Year totals, but birders enjoyed observing the pink pals wherever and whenever they surfaced.

Port Lavaca, Texas, is one area through which the birds cycled, where the pair preferred a shoreline abutting an industrial facility administered by Alcoa. Unfortunately, that multinational aluminum giant prohibits overland access to the shoreline, so it must be approached via Cox Bay, the adjacent tidal body over which the company lays no claim. Were it not for Bob Friedrichs, a local fisherman and birder who frequents the area in his boat, the birds would have gone undetected during their periodic sojourns. Bob kept tabs on the flamingos when they were present, and he graciously shuttled birders to see the flamingo friends when he had the time.

I didn't think American Flamingo was a Big Year possibility until I learned one had appeared in Port Lavaca in late October, apparently in the company of a Greater Flamingo. Following up as I departed the Salton Sea and struggled toward the Rio Grande Valley in the ensuing month, I learned HDNT's story and began thinking about how I might

---

* HDNT was not seen after the end of 2014. Sightings of 492 extended through at least early 2023.

add him to my Big Year list. Mutual contacts furnished me with Bob's phone number, so I called him from Rockport on December 14th.

"Rumor has it you want to see the flamingos," he said.

I replied, "Yeah, that would be great. But I . . ."

"Relax," he interrupted. "I know what's going on. I've been reading your blog for months. I have two kayaks and would be happy to take you out there. When are you going to be in Port Lavaca?"

I replied, "Tomorrow night" (Monday the 15th).

"Perfect. I have Tuesday morning free. Sunrise is just after seven, so let's meet at the eastern end of the Highway 35 bridge at eight. I saw the birds from the motorboat yesterday [the 13th], so they're still around."

Appreciating his generosity and take-charge attitude, I accepted the invitation.

Bob arrived on schedule. In his fifties and fit, he sported a faded pair of pants, a denim jacket, and a sun-bleached baseball cap, a set of squinting eyes peeking out from beneath the tattered brim. We removed the kayaks from the bed of his truck, donned life jackets, and pointed the plastic crafts toward the favored shoreline. The outgoing paddle would be four miles and require a quarter-mile portage midway through it.

My atrophied arms struggled to paddle at the outset, but I established a propulsive rhythm with the aid of a slight tailwind. Bob was very entertaining, and lively conversation carried us around Alcoa's facility and into the surrounding marsh.

Pointing to a sand spit, Bob indicated our portage. We disembarked and dragged our boats across 200 yards of sand and through an equal distance of reeds. It was strenuous, but anticipation built as we resumed our boats and continued.

Unlike my other searches, which required patience and persistence to tease shy birds from concealing habitat, our flamingo foray would be straightforward because five-foot-tall birds couldn't hide in the marshy expanse. Scanning the favored shoreline, I noted two Great Egrets, a Great Blue Heron, four White Ibis, six Willets, and zero flamingos. With no idea where the pair went since Bob sighted them three days prior, we folded our search. It was disappointing, but I decided I wouldn't let an isolated miss prevent me from finishing the year strong. I'd found every species for which I'd searched since leaving the upper Texas coast on April 27th, so I was due for a miss.

The wind had built through the morning, and we needed to steer straight into it to regain our put-in point. Paddling was manageable to the portage point but onerous in the open water beyond it. Waves sloshed over my bow, and my shoulders burned as I struggled to control my kayak. Recovery pauses resulted in maddening backward drift, so I slowed my strokes and thought about something besides my aching arms.

Seizing on the absent flamingos, I couldn't believe the circumstances that allowed two birds—one born in Mexico and the other in Tanzania—to bump into each other in Texas. Given the number of events that could have prevented their association, their friendship seemed an impossible coincidence. Had zookeepers clipped 492's wings, he would never have escaped; had Hurricane Rita not swept over the Yucatán or taken a slightly different course, HDNT might have stayed in Mexico, perished en route, or been blown elsewhere. Theirs was a remarkable association.

———

Early in our relationship, Sonia and I discovered we'd spent years orbiting each other without intersecting. While working at Harvard from 2001 to 2004, I lived on the corner of Harvard and Trowbridge. During that same time, Sonia lived a quarter of a mile from me, on Trowbridge. We frequented many of the same eateries and bars, and I dropped into Grafton Street, the Irish pub where Sonia waited tables, at regular intervals because it was two blocks from my apartment and unavoidable as I walked to the train station in Harvard Square. I was in my early twenties and single, so it's likely I flirted with her on one of my visits. Regardless of the specifics, circumstances didn't facilitate a meaningful connection in Cambridge.

I moved to New York in August of 2004. I lived in the residential tower above the NYU medical center for my first graduate year but moved into my place in the East Village, on 14th Street between First and Second Avenues, in August of 2005. Sonia moved into her place at 12th Street and First Avenue in August of 2006, and—for the next two years—lived 250 yards from me. We again frequented an overlapping set of neighborhood nightspots, and we'd certainly co-occupied Professor Thom's many times before the historic World Series rain delay finally introduced us on October 29, 2008.

By May of 2010, we didn't know how consequential that interaction would prove. I was four months sober, and we were reconnecting after our January split. Neither of us knew where our renewed association was headed, but we knew we had limited time to figure it out before I moved to Boston at the end of the year.

At the beginning of June, Sonia attended a high-profile travel industry event in New York. She called me the following morning.

I answered, "Hey! How was last night?"

"Really good," she said. "And guess what? I won two plane tickets to Australia in a raffle!"

I replied, "So cool! I went six or seven years ago. I spent the whole time partying in hostels, but what I remember was great."

"Well, how would you feel about going back with me?" she asked. "Only catch is that it's gotta happen in the next thirty days. Tickets are use 'em or lose 'em."

My head started spinning. I was in my sixth and final year of graduate school. I'd arranged to start my postdoctoral fellowship at Massachusetts General Hospital in seven months, in January of 2011, and I was working fourteen-hour days to complete a panel of experiments and write my thesis ahead of a late November defense date. Many of my ongoing trials had required lengthy lead-up, so abandoning them midstream could cost me months.

"That would be amazing, but I've got so much going in the lab right now."

Sonia interrupted, "Yeah, I know. Don't make a decision now. I just wanted to plant the idea. It would be a big leap for us, so we should discuss it. Go be a nerd. We'll talk this evening."

I agreed, "Yes, let's do that. Later!"

I hung up and continued to the lab while mulling Sonia's invitation. Scientific considerations aside, an ambitious international trip with Sonia, particularly if poorly organized, could jeopardize the tentative reunification we'd forged. The trip would be a ton of fun, but I was thinking longer term; if I stayed sober and we extended our measured rebuild, then I could imagine Sonia moving to Boston with me. After all, that had been her intention before our January dissolution. I wasn't sure I wanted to risk that enduring vision for a transient vacation.

Sonia and I met at my apartment that evening. Her concerns mirrored mine, but she was optimistic about our ability to manage potential conflicts or difficulties. She explained, "If we can avoid sweating the small stuff, then I think it could be great for us. We had to do Hawaii separately, so this could be a do-over."

She was bringing me around. "It's winter Down Under, so we should probably go north to Queensland, where it'll be warmest," I said.

"Sure, whatever. You plan it. I trust you."

Though nonchalant, her last statement struck me. I'd obliterated her trust in January, so I was thrilled to know the bridge was at least partially rebuilt.

We touched down in Cairns two weeks later. Sonia's decision to leave the planning to me rendered our twelve-day itinerary very birding-focused, but Queensland's national parks facilitated valuable conversation as we searched for honeyeaters, fairy wrens, and cassowaries. The trip went perfectly, an impressive feat given that we camped eleven out of twelve nights, and the uninterrupted access we had to each other allowed gains in emotional and physical intimacy that would have taken months in New York.

Returning to Manhattan in late June, we confronted Sonia's expiring apartment lease in collaboration. With Boston a growing possibility, Sonia was reticent to sign a year-long lease, a consideration that limited her options.

On July 1st, we walked Sonia's minimal possessions from her apartment to mine. She knew roaches put in regular appearances but wasn't ready for the mice, their long-contested insurgency regathering while I was in Australia. I needed to work very long hours during those summer months, and the furry guerrillas seemed to heighten their baseboard offensive whenever Sonia was home alone. She is often loath to show weakness or ask for help, so I found her calls to the lab—"Help! I just saw another mouse!"—as an endearing display of her vulnerability. I don't credit the mice for our reunion, but they offered a barometer to measure Sonia's feelings for me.

I eventually quashed the uprising, and Sonia spoke more of Boston as leaves turned orange. I yearned to know if she would move with me, but I knew she had to guard against the heartache I'd caused her previously. Though I was eight months sober by October, my underlying addiction would never change. Beyond that consideration, Sonia had established a professional presence and made dozens of friends in New York. A Boston move would require her to start over on both fronts, so I avoided salesmanship and gave her space to make whatever decision she thought was best for her.

Watching the San Francisco Giants face the Texas Rangers in the 2010 World Series in late October, that game marking two years since our introduction, Sonia casually announced her decision. Speaking between innings, she said, "We're probably going to need to find an apartment in Boston one of these days."

I grappled for context. "Wait, what? Does that mean you wanna move with me?" I asked disbelievingly.

"Yeah, I'm ready. We're ready," she replied. "We've come so far in the last two years, and I can't wait to see what's in store for us. I think Boston is just the beginning."

I hugged her. "I love you so much, Sonia," I said. "I promise, I'll do whatever I can to make this work. I just want you in my life."

"I feel the same way, Dorian. Whatever happens, we'll have each other. Now get onto Craigslist and find us a place to live. And make sure it doesn't have mice!"

The next month was a whirlwind as I scrambled to finish my remaining experiments. I defended my thesis on November 22nd and spent December cataloging reagents and handing off projects ahead of the holidays. I'd given Sonia no reason to doubt her decision, but I held my breath until she climbed into our rented U-Haul on the snowy morning of January 3rd, 2011.

Assuming the driver's seat, I asked, "Are you ready for this?"

Buckling into the passenger's seat, she confirmed, "I'm ready for us."

On that day, and against very long odds, I departed New York City with my doctorate, 350 days of sobriety, and my best friend, the only woman I ever loved. It was the happiest moment of my life, tears complicating navigation as we left Manhattan, crossed the Bronx, and joined Interstate 95 toward Boston.

"Watch out! What the hell are you doing?" shouted one of the boat captains out the side of his wheelhouse.

Distracted by my reminiscence, I'd allowed my kayak to drift dangerously close to a fleet of commercial trawlers, their nets dragging the bay's bottom for shrimp and shellfish. Intersecting wakes created confused standing waves, and I backpaddled to avoid being clubbed by an extended outrigger. I summoned my remaining strength, escaped the chaos, and reoriented toward the safer course that more-adept Bob had steered.

The return paddle took over two hours and left us exhausted. Bob had seen the flamingos many times, so his effort was entirely for my benefit. Our inability to find the birds was disappointing, but Bob's company and assistance would never be forgotten. I helped him load the kayaks into his truck, thanked him for the assistance, and pedaled toward Palacios, Texas.

Cranking along the coast, my thoughts returned to Sonia. Just as Hurricane Rita had delivered HDNT to Texas and Hurricane Sandy had blown the lapwings to Nantucket, so had the World Series rain delay delivered Sonia to Professor Thom's on the night of our introduction. In life or birding, chance influences everything; as long as I kept my eyes open, opportunity would present, even when least expected. Resolved to control what I could, I prepared to finish the journey that Sonia had put into motion.

THIRTY

Last Call

Relaxing in my Bay City motel on the night of December 17th, I had two weeks remaining in my bicycle Big Year. My original plan had me exploring additional Texas coast before tacking inland, steering toward Dallas, and continuing into Oklahoma, but I wasn't sure I had the energy to execute that ambition after fifty weeks of cycling. With 611 species claimed and only a handful of possibilities remaining, I hoped to push through fatigue and extend my total before the year expired.

Yellow Rail was among those unaccounted few. Bluebird-sized and pear-shaped, the bird is a golden-buff with black streaks on its head, back, and flanks. Yellow Rails nest in marshes in the upper Midwest and southern Canada and winter in the southeastern United States, primarily along the Atlantic and Gulf Coasts. They're unassuming and secretive, and I failed to encounter the species earlier in the year despite traversing much suitable habitat between North Carolina and Texas. Anxious for redemption on my second Texas pass, I'd pushed north along the coast ahead of an invitation to join a Yellow Rail banding outing near Lake Jackson, that one-off scheduled for the night of Thursday, December 18th.

Jennifer, the lead biologist on the project, called on Wednesday, while I was in Bay City. "I assume you're watching the weather?" she asked.

I replied, "Yeah, it looks like two days of thunderstorms starting tomorrow morning. How is this going to work?"

Jennifer explained that the birds shelter during heavy rain. She continued, "We don't want to disturb the rails during that stressful time, and it's dangerous for us to be in the marsh if there's lightning. Tomorrow night looks like a washout, but there's an outside chance of a break on Friday morning. If that happens, then we'll go at five a.m. If not, then we'll wait until Friday night."

The researchers couldn't schedule around me, so I had to disregard the approaching weather and continue to Lake Jackson on Thursday as planned. The storm would make for miserable riding, but I couldn't risk being out of position if the weather broke unexpectedly. The outing would take place in a restricted-access area, and I didn't expect to find Yellow Rail if I missed the opportunity to visit the protected haunt.

The deluge began before sunrise. Water was pooling on the roads by my departure, and gathering traffic made for treacherous riding. An oversized pickup truck heaped waves on me, and I watched a speeding sedan hydroplane for a short stretch, thankfully without incident. By the time I reached the city of Brazoria, streets had been transformed into swamps. A kayak would have been preferable, and I was forced to dismount and push my bike through several shin-deep puddles. Sloshing into my lodging in Lake Jackson midday, I breathed a sigh of relief before cleaning up and hunkering down.

The storm continued through Thursday night and into Friday morning but weakened that afternoon. Conditions holding into the evening, Jennifer instructed me to be at the marsh at nine p.m., the hungry, rain-soaked rails likely to be feeding by that hour.

Joining the four-member team at the appointed time, we unfurled a fifty-foot rope with pebble-filled tin cans attached at five-foot intervals; when stretched between the researchers and dragged across the vegetation, the cacophonous contraption would flush all birds in its path. Furnished with rubber boots, a headlamp, and a long-handled net, I was instructed to walk in front of their phalanx and catch fleeing rails.

I swung the net like a tennis racket and let the trash talk fly. "I'm like John McEnroe," I boasted. "Forehand, backhand, overhead—rails don't stand a chance!"

Incredulous, the team started pulling the rattling rope across the reeds. Clumps of matted vegetation and water-filled depressions made for wobbly walking, but I remained upright while scanning for escapers, the beam of my headlamp sweeping across the marsh like a turret spotlight during a midnight jailbreak.

Ten minutes into our exercise, one of the researchers shouted at me, "There's one! To your left!"

I pivoted only fast enough to see a small bird disappear into the darkness. I didn't even get the net off my shoulder.

"Where were you on that one, Johnny Mac?" teased another team member. "Looks like you're down love-fifteen."

I realized the difficulty of my task but rebuffed the ribbing with more braggadocio. "I'm all over the next one. Take it to the bank!" I said.

The opportunity to back-up my bombast availed when a second rail flushed to my right a few minutes later. I had a clear view as the bird passed through the beam of my headlamp, but the damn thing must have activated its hyperdrive because my full-extension flop netted only air. Soaked from head to toe, I expelled a mouthful of marshy water and plucked a reed from my teeth. I stood and spoke, "This should be an Olympic sport!"

"Everyone underestimates them," said a third researcher. "They're out in numbers tonight, so you'll get one eventually—we hope."

They resumed dragging, a third example lifting out of the marsh fifteen minutes later. I closed the distance with a series of strides, but the terrified bird made an abrupt about-face and put down in front of me as I readied to take my pursuit airborne.

Crap! I'm going to land right on the stupid thing! Abort! Abort!

I slammed on the brakes, lost my footing, and went down flailing; my body crumpled like an accordion as I hit the ground. Wiping water from my face, I saw the rail peering at me from three feet away and quickly lowered my net over the quizzical quarry. It was the ultimate triumph of man over beast—or so it felt at that soggy, midnight moment.

The team helped me up and removed the rail from my net. They weighed it, noted physical characteristics, and affixed a tiny metal band to one leg, those combined processes requiring just ten minutes.* When the bird was ready for release, Jennifer indicated I should do the honor. She showed me how to grasp the delicate body, and I took a few moments to examine the stout yellow beak, scaly legs and feet, and subtle plumage features.

Stroking the golden breast with my finger while beady eyes returned my gaze, birds seemed the stuff of dreams. Evolved in endless sizes, shapes, and colors, they occupy every habitat on Earth, even those too harsh for humans. Their behaviors are as intriguing as their evolutionary histories are fascinating, and their ability to fly leaves us yearning to experience the world from their heavenly perspective. Even flightless varieties inspire joy: penguins, ostriches, cassowaries, and kiwis among those curious but comical outliers. Just one of 10,000-some bird species, the Yellow Rail occupies a little-read chapter in the ornithological annals. I'd not encountered the bird at any point in my birding career, and I doubted I'd be afforded such a close view again.

Holding the captive, I closed my eyes; without vision to suggest the bird's presence, my hand felt empty. That such a delicate creature could migrate from Canada to Texas and back each year was astounding. Between birds' physical abilities and their navigational talents, I didn't know which was more impressive. I didn't want to deny the rail either for any longer, so I opened my eyes and uncurled my fingers, our encounter ending as the bird departed my hand.

Species #612 netted, I retired to the small room I'd secured atop the local bait-and-tackle shop. Though fishy, the hovel saved me from returning to central Lake Jackson at a Friday night hour when bars would be closing. If placed in the path of a speeding drunk driver, my body would be as comparatively frail as the rail's.

I birded around Lake Jackson on Saturday and used Sunday to participate in the Freeport installment of the National Audubon Society's

---

* Metal bands are now yielding to GPS transmitters because electronics return positional data in real time and without the need for recovery. As smaller transmitters are manufactured, the body-size threshold for electronic tags decreases.

Christmas Bird Count, a nationwide census that has surveyed wintering birds since 1900. There is friendly competition between participating municipalities, Freeport annually among the top counts in the country, and the pooled data is a valuable snapshot of bird populations from one year to the next. I volunteered to count all individuals of all species I observed from Freeport's north jetty, a vantage that satisfied two ulterior motives. The first was my desire to see Black-legged Kittiwake, a high-latitude gull I missed in New England. While I wasn't likely to find the bird, with only a few recorded on the Texas coast each winter, my chances were better on the jetty than anywhere else in the count area.

Beyond spotting the kittiwake, Kenn Kaufman, one of my childhood birding idols, had accepted the same assignment during his 1973 Big Year. Then just nineteen years old, Kenn hitchhiked 70,000 miles around the United States and Canada and tallied 666 species, an incredible total in that era and on his shoestring budget; he famously consumed cat food at penniless points. Winding his adventure down at Freeport, he went for a surprise swim when a rogue wave knocked him off the north jetty. He survived the plunge and later chronicled his hitchhiking-birding adventure in *Kingbird Highway*, a transcontinental tale that inspired my adolescent Big Year dreams before alcohol swept them out of my conscience. I didn't observe the kittiwake during my seven-hour vigil, but my time on the jetty lent the envisioned connection to Kenn. In a Big Year landscape dominated by wealth and petroleum, I smiled when I imagined myself alongside one of my idols; on his budget or on my bicycle, no Big Year birder had done more with less.

My coastal birding concluded, I pedaled out of Freeport on December 22nd. I followed Highway 36 to Rosenberg, crossed the Brazos River, and continued to Monaville. Harris's Sparrow (#613) was an effortless addition on my host's feeder there, but northern headwinds extracted payment across the ensuing two days; billowing gusts allowed just thirty-seven miles to Navasota on the 23rd and another fifty to Hearne via College Station on the 24th. Tailwinds were an apropos Christmas gift. Those pushed me 112 miles to Ennis, and I reached the Dallas suburbs on the 26th. Arriving at White Rock Lake that afternoon, I intersected an extralimital Little Gull, which had been reported for the previous week. I missed the bird on my northeastern leg, so the bird was a fortuitous salvage at year's end. Species #614 secured, I retired to a nearby motel to contemplate my final moves.

With five days remaining and only two unclaimed species within a week's ride of Dallas, I considered mutually exclusive options. The first was to prioritize Smith's Longspur, a songbird that nests on arctic tundra and winters on the Great Plains. A flock of thirty had been observed east of Dallas at Lake Tawakoni for several days, and the birds seemed all but guaranteed if I covered the sixty miles to the abandoned airstrip they'd favored since their discovery. Unfortunately, riding to Tawakoni wouldn't leave time to return to Dallas, continue northwest into the Oklahoma Panhandle, and search for the other bird, Lesser Prairie-Chicken. My only shot at that range-restricted species would be to skip the "sure-thing" longspurs at Tawakoni and proceed directly to Oklahoma, a distance of 350 miles. It would be a monumental effort at year's end, but the coincidental chance of intersecting the wider-ranging longspur en route suggested a storybook ending where I found both birds to finish the year in a blaze of glory.

My Inner Addict seized on that narrative. "You gotta go to Oklahoma!" he shouted. "It'll be a huge high if you get both birds!"

My Sober Self responded, "You really want to go all-in this late in the year? You're more likely to miss both birds than see either given the huge distance and limited daylight. Settle for the Tawakoni longspurs instead."

Inner Addict countered, "When did you do the safe and easy thing this year? Getting the longspurs and leaving the final four days of the year as dead air is chickenshit, like leaving a bar before last call."

The Addict was right. Whether detouring to eastern Colorado for Greater Prairie-Chicken, extending to Washington State for Spruce Grouse, Boreal Chickadee, and Gray-crowned Rosy-Finch, or backtracking through California for Pacific Golden-Plover and Rufous-backed Robin, I'd reaped rewards every time I pushed myself.

Sober Self pleaded, "You've survived an impossible journey. One more bird won't change what you've accomplished. And what about Sonia? You promised her you'd do everything you could to return safely. Neither of you needs this binge."

Alone in my motel room, enough was finally enough. I'd thought much about idealized projections of the future during my travels and realized I needed to focus on enjoying what I already had. Bike-birding had supplied structure and purpose across the year, but Sonia would be my companion and lighthouse long beyond the expiration of my adventure. She'd precipitated my sobriety and motivated my escape from career confusion, and it wasn't worth risking a lifetime with her to hold onto an undertaking that would expire in five days. Life would go on without the prairie-chicken—unless I was run over while looking for it.

Departing the Big Year bar before last call, I initiated the sobering process with a relaxed, five-hour ride to Tawakoni on the 27th. I could have pushed to find the longspurs that afternoon, but wind and rain suggested I shelter at a motel and enjoy the breathing room I'd afforded myself, that pause validated when the anticipated flock flew over my head at the runway the following morning. My backlit view of the brown birds was marginal, their rattling flight calls more diagnostic than visual field marks, but the encounter was enough to include the Smith's as species #615. My emotions were subdued compared to an imagined prairie-chicken intersection, but the longspur was the modest triumph I needed at that end-of-year juncture.*

My legs and the birding possibilities exhausted, I retreated toward Dallas, where Kate, a close friend from Hotchkiss, had invited me to ride out the year's final days with her family. I liked what I could remember of her husband, Kurt, from their 2005 wedding—I was blacked-out for

---

* The American Birding Association made Egyptian Goose, an introduced species, countable midway through 2014. I didn't assign the species a number when I saw it in Florida in March, so I designated it #616 at year's end because species made countable during (but not after) a Big Year can be counted toward the total. Florida's introduced Muscovy Ducks were count-able when I observed them, but I forwent numbering the species at that time in case I intersected a native example in Texas. That didn't happen, so I assigned Muscovy #617 at year's end. The American Birding Association accepted our Red-legged Honeycreeper sighting in July of 2016. I designated it #618 at that time.

most of the two-day event—and I was overdue to meet their two sons, ages four and eight. I wasn't keen to ride the entire ninety miles to their Lewisville house post-longspur, so I ducked into a hotel near Greenville after completing the first thirty. As long as I survived the final sixty, I'd be able to experience life beyond the bicycle.

I showered, ate dinner, and collapsed in front of the television. The weather report suggested my decision to skip the prairie-chicken was wise; with wind and freezing rain to batter Oklahoma and northern Texas the following afternoon, I would have been stopped short of the bird's range. Thankful to have avoided that disappointment, I settled into college football while awaiting the year's most anticipated communication.

THIRTY-ONE

Connecting the Dots

My phone rang at eight thirty p.m. I knew who it was and dispensed with pleasantries.

"Are you here?" I asked desperately.

"Yep! I'm in the lobby," she replied.

"OK, sit tight. I'll come get you."

I threw on my shirt and shoes and scampered down the stairs rather than wait for the elevator. Throwing open a steel fire door, I saw Sonia, her smile outshining the lobby Christmas tree. I ran to her and melted into her arms, her lips meeting mine as her hair swept across the side of my face.

"How do you feel now that it's over?" Sonia asked.

"It's been incredible, but I'm tapped out," I said. "The riding was tough, but the logistics might have been tougher. There was so much to think about each day: birds, routes, road conditions, traffic, weather, food, water, lodging, blogging. It never ended."

"I know you're not good at sitting still, but you have to give yourself some downtime after this," Sonia suggested. "You need it even if you don't realize it."

"Yeah, I know. And I know it's going to take a while for this whole thing to sink in. I'm too close right now. It's kinda like my drinking. I had to get beyond it before I could understand it. Does that make sense?"

"Totally. And it's great to hear how you're thinking. This is exactly the sort of change I hoped you'd experience. You were obsessed with controlling the future before you left, but maybe now you can relax a little. It's OK to stop and just be happy."

She was right; her unyielding belief that everything would work out somehow, someway, was one of her most endearing and gravitational qualities.

"You know," I said, "I think being apart has made us closer. There were a lot of times when I questioned what I was doing on the bike, but I never worried about us."

I couldn't imagine my life had it not intersected hers. We'd experienced every emotion in the intervening six years—two of those spent in New York, three in Boston, and the most recent apart—and I was ecstatic our pair-bond had strengthened when it could have weakened.

I gave Sonia a hug and hit the road at eight a.m. the next morning, the 29th. My hope was to reach Lewisville before the weather deteriorated. I texted Sonia when I was fifteen miles from Kate's house. Sonia departed our hotel at that time, and we rendezvoused in Kate's driveway an hour later, exactly as northwest wind and freezing rain materialized. Relieved to have survived the year's crowning ride, I rang the bell to the two-story brick abode.

Kate was one of my partners in crime through high school and college. She had a party streak across those years but, like most, slowed through her mid-twenties; while she was marrying at age twenty-seven, I was still expanding my narcotic resume and drinking myself into oblivion. Despite that behavioral divergence, she didn't levy judgment on or lose patience with me as did some of our mutual friends. Kate always took me as I came and was one of the most loyal people I knew.

Kate answered the door. "You made it!" she said. "And just ahead of the storm!" She turned toward my better half and continued, "And you must be Sonia! I didn't think there was a woman on Earth who could straighten this guy out! It's great to finally meet you!"

Kate instructed, "Wheel your bike around to the garage. The guest room is upstairs to the left. Towels on the bed. Sonia, come with me. Let's crack a bottle of wine while Stinky gets cleaned-up!"

"OMG! You're my new favorite! Lead the way!" Sonia replied.

I dealt with my bike, dragged our stuff upstairs, took a shower, and joined the ladies in the living room.

"What's your plan for the next few days?" Kate asked. "You're welcome to stay as long as you want."

"I think we'll hang here until the first. There's no rush for us to get to Philly to see my parents, and I'd like to finish the year petroleum-free, even if it means loafing around here for the last two days."

"No problem," Kate said. "We're gonna have six or seven couples and their kids over for New Year's, so you should join us. Otherwise, just come and go as you want."

Between sleeping late, socializing, watching college football, and writing several final blog entries, the next two days flew by. The New Year's gathering was approaching, and Sonia and I offered to help ready the house on the afternoon of the 31st. Tasked with organizing the kids' toys while Sonia lent assistance in the kitchen, I picked up a drawing book, the sort where numbered dots are connected to reveal the image of a rabbit, truck, or mermaid. I possessed zero artistic ability as a kid, so I enjoyed the rudimentary exercises in what little of my youth wasn't spent looking at birds or throwing rocks at trains. Feeling nostalgic for my childhood, I opened the book, images of lumpy lambs and misshapen mushrooms igniting my imagination as I leafed through the pages.

I suddenly pictured my bicycle Big Year as a similar exercise. Snowy Owl in Massachusetts was Dot #1; Snail Kite in Florida was Dot #213; Blue-throated Hummingbird in Arizona was Dot #396; Greater Sage-Grouse in Colorado was Dot #503; Gray-crowned Rosy-Finch in Washington was Dot #522; Pacific Golden-Plover in California was Dot #563, and Smith's Longspur in Texas was Dot #615. With the provisional honeycreeper included, my total of 618 exceeded anything I thought possible when I wobbled into the Salisbury darkness a year earlier.

Reaching for my wallet, I removed and unfolded the paper map that had guided me since my departure. Initially an impossible aspiration, the convoluted blue trace had crystallized into an unlikely reality across the ensuing twelve months. Solitary headspace had suggested that the road to personal understanding is neither smooth nor straight, and the completed journey was a representation of what I could accomplish with a crazy idea and unyielding determination.

Staring at the map, lost on the trace of my recent past, I realized that my life was the sum of the circumstances I inherited and the decisions I made. I didn't choose to be an alcoholic, my family history and genetics tending me in that direction, but drinking on any given day was a conscious decision, even if I didn't understand it in the moment. Sonia didn't leave me because I was an alcoholic; she left me because I continued to choose alcohol and drugs over her.

I was as much an alcoholic on December 31st, 2014, as I was on January 18th, 2010—the date of my last drink—but I'd abstained across the intervening five years because I realized the alternative would be disastrous. My desire for annihilation hadn't waned, and I knew that the line from my first drink to the twentieth was still perfectly straight. I had, however, found substitute highs in the absence of alcohol, and my life-long inability to relax made sense after undertaking an 18,000-mile bicycle trip without prior cycling experience. I needed to overdo everything to feel anything; to me, moderation didn't make any more sense in the saddle than it did at the bar or in the research laboratory.

That revelation illuminated a view beyond the bicycle because I finally understood that my addictive tendencies were transferrable. Alcohol was a convenient deployment for fourteen years, but I possessed the power to repurpose my condition to productive ends. Rather than denying my affliction, I would own it. I would embrace it. And I would nurture it; if I put the energy I once put into drinking and drugging into other channels and projects, then there was no limit to what I could accomplish. I didn't know what the future looked like, but I knew that I was better prepared to face it after a year-long bike ride than if I had done anything else with the time. At the end of it all, the birds were just the dots I connected to draw a happier picture of myself. Folding up my map, I couldn't wait to see what I would do next.

Party guests started arriving at eight p.m. I mingled and raised a token glass of orange juice as the countdown began. Ten seconds later— and while planting a huge kiss on Sonia—the curtain came down on 2014. My Big Year was over, but my journey would continue.

Epilogue

I n January of 2015, three weeks removed from my bicycle Big Year, Sonia's mother, Yolanda, was diagnosed with stage 4 colon cancer and given less than two years to live. Though in the process of returning to Boston, we immediately reoriented toward Los Angeles so Sonia could spend time with Yolanda. A friend and postdoctoral colleague from Mass General, Carolyn, had recently secured a faculty position at the University of Southern California. When I told her about our impromptu move, she invited me to join her research group. I wasn't thrilled about returning to academic science, but it was a perfect arrangement at a moment of crisis; I found purpose and income while we spent time with Sonia's family, and Carolyn secured a proficient experimentalist who could help train her students.

Sonia and I didn't talk of marriage before moving Los Angeles; kids weren't in our future, and we weren't motivated to codify our love in traditional ways. However, my thinking evolved as her mother's condition deteriorated. If we were going to get married, I felt it should be while Yolanda was alive. I hadn't made any concrete plans by October of 2015, when I drove Sonia and Yolanda north to Napa for a four-day getaway, but the moment seemed right to me as we returned south along the coast. Standing on Point Piños in Pacific Grove, the waves crashing on the rocks below us, I spontaneously dropped to one knee and, while holding Sonia's and Yolanda's hands, asked Sonia to marry me. She said yes, the lack of a ring insignificant beside her ailing mother's joyous tears.

Sonia and I married in May of 2016. Not fans of pomp or circumstance, we held the wedding at a local roller rink and requested that guests attend on skates and in costumes. Sonia wore a lime-and-chartreuse mod-type dress while I sported a white leisure suit, and the event was a rousing success with the likes of Batman, Super Mario, Minnie Mouse, and Betty Boop in attendance. Yolanda was too frail to skate, so friends and family took turns pushing her around the rink in her wheelchair. Yolanda passed in November of 2016. Sonia and I moved to the Bay Area in May of 2017, when Sonia accepted a position at Airbnb.

Sonia has remained my rock throughout, but my professional life has continued to see significant flux—all for the better. Speaking invitations poured in as soon as my Big Year ended, first from birding clubs and later from birding festivals around the country. More surprising were the high schools and universities that wanted to hear from me, my life choices resonating more than the birds in those educational settings. Carolyn granted me latitude to pursue such bird-related opportunities, and the underpinnings of an alternate career took shape through 2015.

That process accelerated when I was pulled into a multiagency birding and conservation project in Colombia in 2016. I made two trips to the country that year and, as a result of that work, secured a side hustle as an ecotourism correspondent for the Nature Travel Network. Subsequent experiences in Belize, Guatemala, Spain, Honduras, and Taiwan fueled my interest in ecotourism, and I was offered a dream consulting gig with the National Audubon Society in 2018. My assignment was to spend two months in Colombia and help the government identify areas with conservation and ecotourism potential. Because many neotropical migrants nest in the United States and winter in Colombia, the initiative is expected to pay international dividends.

In 2019, I began leading birding tours. The pandemic guaranteed a slow start, but I am now very busy. Beyond a host of North American destinations, I am currently spending a lot of time in Africa; between copyediting and publication of this book, I will visit South Africa for the third and fourth times, Namibia for the second time, and Botswana and Kenya for the first times. It is wonderful to share my love for birds, animals, and photography with clients, and I cannot wait to see where on Earth my avian fascination and associated wanderlust will take me.

EPILOGUE

I know firsthand how difficult it is to make proactive personal change, but I hope my story illustrates the benefits of accepting that challenge. While my decision to get sober was paramount, my emotional and professional metamorphosis would have been incomplete had I not subsequently departed academia, that pivot opening the door to the bicycle trip and the opportunities that have flowed from it. In both instances, I overcame familiarity, fear of failure, and perceived loss of identity and discovered that I have control of my future, even if insecurity, circumstance, and practicality sometimes suggest otherwise. In short, I got out of my own way and allowed sobriety and the bicycle to chart an alternate course, one I couldn't imagine under the influence of alcohol or inside the confines of academia. Most important, I am happy, and that's not something I thought possible at my first AA meeting or the day my postdoctoral project cratered. Trust yourself, follow your passion, and forget about the rest. It'll work out somehow, probably in ways you never anticipated.

Acknowledgments

There aren't enough words to acknowledge everyone who supported my bicycle Big Year. Between bird-finding assistance, lodging, and emotional support, I cannot thank the birding community and the American public enough. Although I had to omit or gloss over many of the interactions I had along my arc, which I regret, every interaction, conversation, and note was meaningful, and I would not have accomplished half of what I did without the unyielding and unconditional backing I received. The United States is often painted as a cold and hostile place, but my two-wheeled travels revealed the opposite. Whenever I needed assistance, there was someone there—academic or evangelical, rich or poor, gay or straight, White or Black, helping hands were always extended. And, at a time of political division, I think that's worth celebrating.

Beyond my general gratitude, I'd like to mention one person and several organizations. In 2009, while I was at NYU, I met Tom while chasing the same rare bird in upstate New York. He was cynical and funny, lived only a few blocks from me in Manhattan's East Village, and claimed a decade of sobriety at our intersection. We shared much in common, and he served as a valuable confidant—and my de facto sponsor—when I confronted my own addiction in 2010. I could have written chapters about our string of weekend birding outings, but I chose to keep the story focused on Sonia and me. Tom and I are still in contact, and I include him among my closest friends.

In addition, I would like to thank Warmshowers, an online community of traveling cyclists who house and feed each other. I spent more than 100 nights with Warmshowers hosts, and their generosity provided needed interaction while minimizing lodging costs.

I also want to thank Best Western. The brand sponsored me with $6,000 to use at their hotels (25 percent of my total budget, bike and gear included), and the nationwide chain was my go-to whenever I couldn't connect with a friend, birder, or Warmshowers host. I spent 54 nights with Best Western, and I knew I was in for great sleep and free breakfast whenever I rolled up to one of their properties.

Hunt's Photo and Video was another valued partner. Beyond furnishing me with a compact carbon fiber tripod to support my scope, the New England–based retailer set up periodic deals for my blog readers. At the year's conclusion, Hunt's exhibited photos from my trip in their gallery in Melrose, Massachusetts. Their customer service is fantastic, and they are my preferred source whenever I can't find whatever photo gear I want on the used market.

Through fundraising during my Big Year, I raised nearly $50,000 for bird conservation and programming. The proceeds were split between the Conservation Fund (80 percent) and the American Birding Association (20 percent). Victor Emanuel Nature Tours, the organization that runs Camp Chiricahua, kindly matched the $2,500 that blog readers donated during a critical two-week window. Their timely involvement helped breathe energy into that aspect of the project.

Massive thanks to everyone who read my blog, bought me lunch, sent notes of encouragement, put me up for the night, helped me find birds, or took a casual interest in my adventure. At the end of it all, the people were the best part of my journey.

About the Author

Dorian Anderson Photography

An avid birder since childhood, Dorian Anderson abandoned his hobby at age fifteen, focusing instead on a demanding scientific career while simultaneously struggling with substance abuse. He earned a degree in molecular and cellular biology from Stanford, conducted predoctoral research in molecular embryology at Harvard, and earned his doctorate in developmental genetics and molecular cell biology at New York University. While working as a postdoctoral fellow at Massachusetts General Hospital, Dorian decided to leave the academic rat race and focus on the next phase of his life, returning to birding. In 2014, he embarked on his Biking for Birds project, the first North American Big Year completed entirely by bicycle. During this incredible journey, he biked 17,830 miles and observed 618 bird species while raising funds for bird habitat conservation. Since his cycling Big Year, he has transitioned to a professional life as a birding guide, writer, and public speaker.